THE TRAGIC END OF THE COAST GUARD CUTTERS WHITE ALDER CUYAHOGA & BLACKTHORN

By

Edward Leo Semler Jr.

Copyright © 2024 by Edward Leo Semler Jr.

All rights reserved by the author.

First Edition: 2024

Library of Congress Control Number: 2024905053

ISBN: 978-1-737-6472-5-6

Printed in the United States of America

City of Publication: Schulenburg, Texas

Cover layout by Edward Leo Semler Jr.

To the crews of the White Alder, Cuyahoga, Blackthorn, and their family's

TABLE OF CONTENTS

White Alder	1
The 7th of December 1968	19
Aftermath	33
Cuyahoga	55
The 20th of October 1978	73
Aftermath	91
Blackthorn	129
The 28th of January 1980	149
Aftermath	179
Closing	215
References	217
About The Author	221

WHITE ALDER

One of the main functions of the United States Coast Guard is maintaining aids to navigations on Americas waterways. And one of the more important areas of their responsibility is that of the Mississippi River.

I participated in this endeavor while serving in the Coast Guard Cutter Sumac (WLR-311) in the mid 1980's. She was homeported out of St. Louis, Missouri, and responsible for maintaining the aids to navigation down to Cairo, Illinois where the Ohio River empties into the Mississippi River. Dark and murky, the Mighty Mississippi can reach speeds of 3 mph. This may seem slow, but for water, it's fast.

Ensuring that Americas waterways are safe for travel may seem like a mundane and relatively safe task, but this can be very dangerous. And as in the story you are about to read, deadly.

The risks imposed on a vessel working the aids to navigation are primarily operating in opposing traffic patterns and in a condition of restricted maneuverability. And what does that mean?

Aids to navigation come in several forms but are primarily buoys. These buoys mark the waterway like stripes on a highway on land, with a left – port, and right – starboard, boundary along with a direction of travel. These can sometimes also be in the form of colored boards or markers placed on the bank of the river or inlet.

To maintain these aids, the maintenance vessel occasionally must operate in opposing traffic patterns, which can be extremely dangerous. Sort of like a work crew on a land highway. And because the maintenance vessel is conducting work, such as setting or moving a buoy, they are limited in their maneuverability to avoid other vessels.

And just like rules for driving a car on land, there are rules for operating a vessel on waterways, and these have been internationally adopted. So, unlike driving on land, where there are differences in driving in let's say the United States compared to Great Brittian, rules for operating on water are universally governed under one set of rules or laws. Oddly enough these are referred to as the "Rules of the Road" – not "Rules of the Water." They are further broken into International and Inland Rules, depending on whether you are operating on the open ocean or inland water ways such as a river.

These rules designate the proper way to transit, who has the right of way, what markings must be displayed during the day, and what lights must be displayed at night. To operate military and commercial vessels on waterways worldwide you must pass examinations on these rules.

I'm laying out this information to prepare you for the stories you are about to read. Because when vessels collide, there has to be a reason. At the scene of any accident there is a gathering of facts. And those facts are then compared to a set of rules or laws. And the outcome of these findings determines what rules or laws were broken to cause the accident.

The incident you are about to read happened the 7th of December 1968 on the Mississippi River between the United States Coast Guard Cutter White Alder (WLM-541) and the Chinese owned S.S. Helena. The aftermath was the total loss of the White Alder and 17 of her 20-man crew.

After the incident there was an extensive Marine Casualty Report[1] which I will outline in the following chapters. This report included findings by the National Transportation Safety Board – Department of Transportation and the U.S. Coast Guard.

In 1968 all the branches of the Armed Forces – Army, Navy, Marine Corps, and Air Force, served under the Department of Defense except one – the Coast Guard. The Coast Guard had been moved the previous year, 1967, from the Department of the Treasury to the Department of Transportation.

Since the White Alder was therefore a vessel under the control of the Department of Transportation, and the Department of Transportation was the governing body of the initial investigation into the collision between the White Alder and Helena, there was or could have been, a lack of transparency. Typically, the governing body overseeing an investigation would be a third party. The Department of Transportation recognized this and rightfully addresses it in their report.

As we dive into this story, the Marine Casualty Report[1] will lay out the events of the incident and I will add newspaper reports and other supporting documents to help broaden the picture of what is going on.

So, let's start by getting to know these vessels and their crews.

U.S.C.G.C. White Alder (WLM-541)

United States Coast Guard Cutter White Alder began her life as a United States Navy "lighter" vessel. This type of ship was designed to carry cargo, such as ammunition, from shore to deep draft vessels. She was commissioned in 1943 as the YF-417 and was transferred to the Coast Guard in 1947 to be converted into a coastal buoy tender. After conversion she was given the hull number WLM-541. She was 132' in length, had a breath of 30' 9", had two propellers and was powered by two diesel engines. The vessel was considered very maneuverable.

After the Coast Guard converted her, she was stationed in New Orleans, Louisiana where she would spend her entire career. Her primary duty was that of aids to navigation but was occasionally tasked with search & rescue and law enforcement.

Her crew generally consisted of 1 chief warrant officer (CWO) and 20 enlisted men. On the day of the collision with S.S. Helena she had a compliment of 1 chief warrant officer & 19 enlisted men. One of the assigned crew members, Boatswain Mate Chief (BMC) Richard Batista, was left behind in New Orleans on leave.

The crew aboard on the 7th of December 1968 was as follows:

CWO Samuel C. Brown Jr.[2]

Chief Warrant Officer (CWO) Samuel C. Brown Jr. – 41 of 1608 Gaff Road Chesapeake, Virginia. CWO Brown was the commanding officer. He was a native of Bloomington, Illinois, where he was born on the 31st of October 1927. His parents moved to California when he was six months old, and he joined the Coast Guard when he was 17. His mother eventually moved back to Bloomington, and he visited her there regularly. On one of these visits, he met and married Freda Nuss. They had 4 children, 2 boys and 2 girls who resided in Chesapeake, Virginia. He also had a stepson and stepdaughter which I presume were from Freda's previous marriage.

Chief Engineman (ENC) William J. Vitt – 29 of Mereaux, Louisiana. He was a native of Louisville, Kentucky and married to Barbara. They had 4 children, 2 boys and 2 girls.

Engineman First Class (EN1) John B. Rollinson – 27 of Buxton, North Carolina. He enlisted in the Coast Guard in February of 1964. He was married to Vera Mae Thaxton, and they had 2 sons.

Electricians Mate Second Class (EM2) Michael R. Agnew – 24 of Miami, Florida. He was married to Eileen Rabinowtz. He also had a brother, Daniel A. Agnew, serving with the Coast Guard at a Light Station on Dry Tortugas Island, Florida.

Quartermaster Second Class (QM2) John R. Cooper Jr. – 32 of 2823 West Kirby Street Tampa, Florida

Commissaryman Second Class (CS2) Charles R. Morrison Jr. – 30 of Clarksville, Tennessee, formerly of Louisville, Kentucky. He enlisted in the Coast Guard in March of 1966. He was married to June Rita Rodriguez, and they had 1 son.

Yeoman Second Class Joseph A.R. Morin[6]

Yeoman Second Class (YN2) Joseph A. R. Morin – 32 of 106 Chestnut Street Lewiston, Maine. He had joined the Coast Guard when he was 17 but transferred to the Army and then the Air force before settling again with the Coast Guard. He saw action with the Coast Guard in Vietnam and participated in Operation Deep Freeze with the Coast Guard in 1955 to 1956. He was planning on getting married in the summer to Anne Lee, a nurse in New Orleans.[4]

Boatswain Mate Second Class Richard Krauss[3]

Boatswain Mate Second Class (BM2) Richard Kraus – 23 of St. Petersburg, Florida. He enlisted in the Coast Guard in 1965 after attending Boca Ciega High School and St. Petersburg Junior College. He was Married to Chris Mudry. He also had a 21-year-old brother in the Coast Guard stationed at the Naval Air Training Center in Memphis, Tennessee.

Boatswain Mate Third Class (BM3) Guy T. Wood – 24 of Willis, Virginia. He enlisted in the Coast Guard in August of 1965.

Engineman Third Class Walton E. O'Quinn[7]

Engineman Third Class (EN3) Walton E. O'Quinn – 37 of Brunswick, Georgia. Was married to Virginia Morales and they had a daughter. He had 15 years of military service. It's not known if it was all in the Coast Guard.

Fireman (FN) Maurice Cason – 19 of 2921 Jasmine Street Denver, Colorado. Was married to Sharon and the son of Air Force Master Sergeant Alfanso Carson stationed at Lowry Air Force Base in Colorado.

Fireman Bruce L. Kopowski[9]

Fireman (FN) Bruce L. Kopowski – 22 of Kent, Ohio.

Seaman (SN) Richard W. Duncan – 20 of Oakdale, California. A native of Dalhart, Texas, he moved to Oakdale when he was three years old. He joined the Coast Guard when he was a junior at Oakdale Union High School. He attended basic training in Alameda, California and then was stationed in Hawaii before serving in Vietnam. He received his high school diploma while in the Coast Guard and had been assigned to White Alder for about a year.[5]

Seaman (SN) Steven D. Lundquist – 20 of 2557 Lavender Street Los Angeles, California. He graduated from Eagle Rock High School in 1966 and was married to Elaine.

Seaman Frank P. Campisano III

Seaman (SN) Frank P. Campisano III – 20 of New Orleans, Louisiana. He enlisted in the Coast Guard in June of 1967.

Seaman Apprentice (SA) Larry V. Fregia – 19 of Conroe, Texas. He enlisted in the Coast Guard in June of 1968.

Seaman Apprentice (SA) Roger R. Jacks – 20 of Irving, Texas. He enlisted in the Coast Guard in March of 1967

Seaman Apprentice (SA) Ramon J. Gutierrez – 21 of New Orleans, Louisiana.

Seaman Apprentice (SA) Walter P. Abbott III – 24 years old of 312 South Union Natchez, Mississippi. He was born in Natchez and attended Southwest Junior College at Summit, University of Mississippi and received a Bachelor of Science degree from Delta State College. In May of 1968 he enlisted in the Coast Guard.

Seaman Apprentice (SA) Lawrence E. Miller – 23 of Abbeview Avenue Willow Grove, Pennsylvania

The following drawings are of White Alder from a starboard side profile and a top view main deck profile.[23]

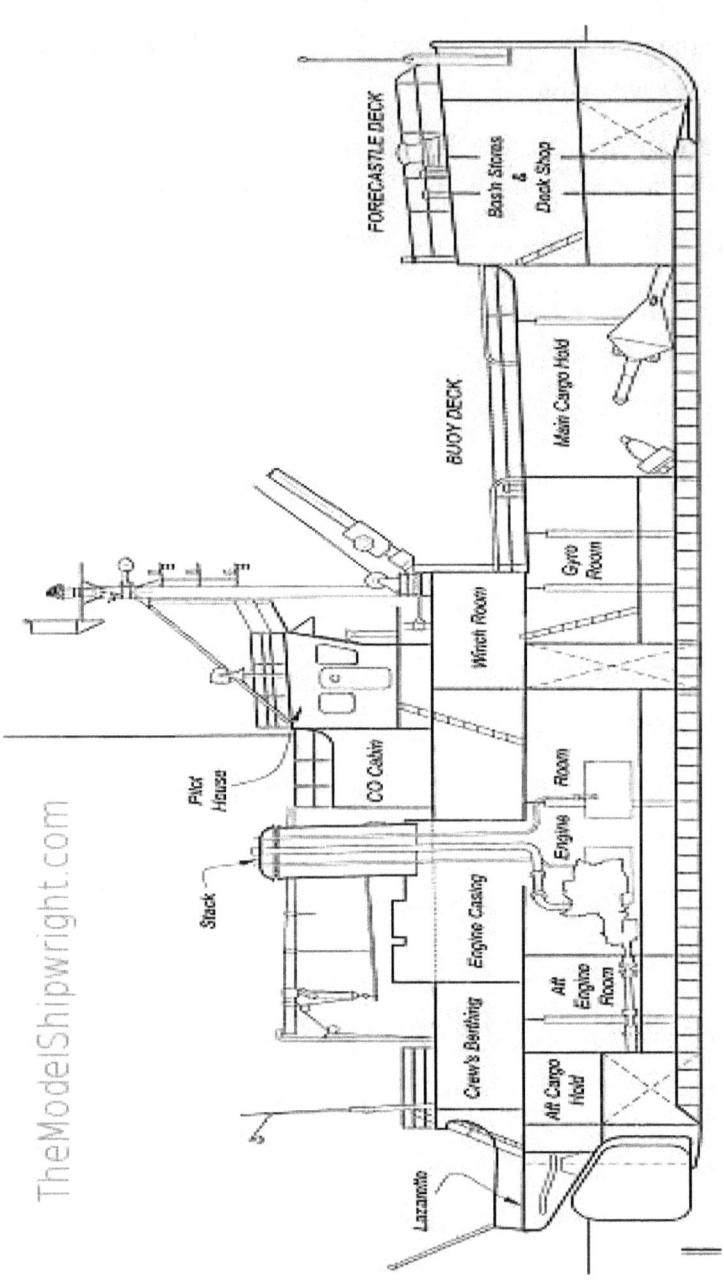

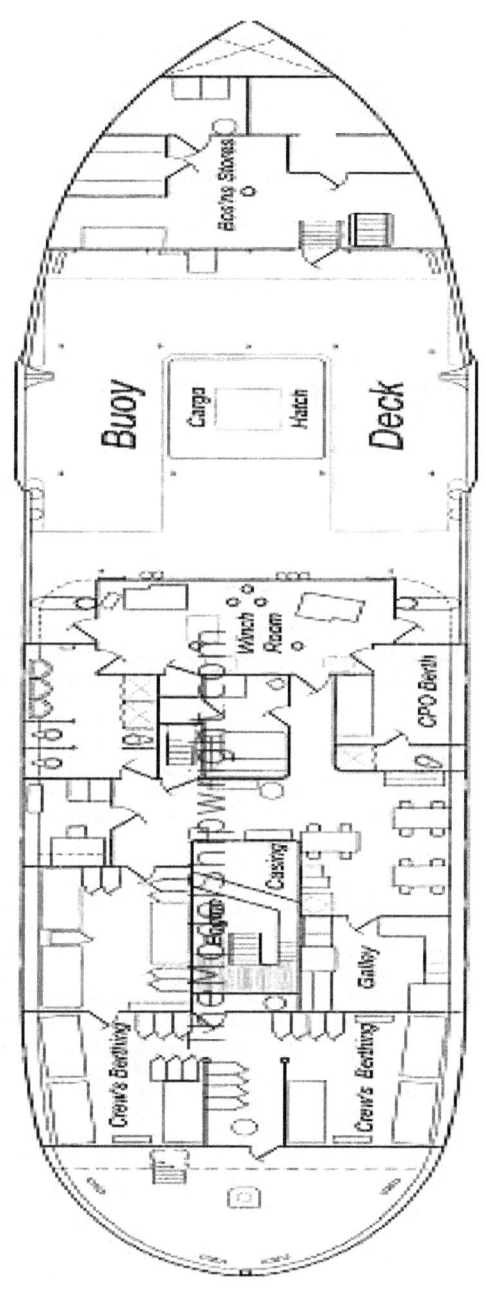

Typical U.S. Victory Ship & what S.S. Helena would have looked like

The S.S. Helena started her life as a United States Victory Ship. This class of cargo ship was produced in mass quantities during World War II. She was built in 1944 and had a length of 455' with a breadth of 62'. She had a single propeller and was powered by steam boilers. At some point she was sold to China and in 1968 was operating as the S.S. Helena #130.

The captain and crew were all Chinese residents of Taiwan. Their names were redacted in the Coast Guard and Department of Transportation reports. However, I was able to find some listed in newspaper articles covering the investigation.

The known crew aboard on the 7th of December 1968 was as follows:

Shih Teh-Chang - Captain

Herold Rowbatham - Pilot

Wang Wei-Sheu – Third Mate

Chu Ching-Ling – Lookout

Wu Ko-Jen - Seaman

THE 7th OF DECEMBER

On the 7th of December 1968, the weather conditions near Bayou Goula, Louisiana was a cool 48°F. The winds blowing at 20 mph and coming out of the northeast made it feel even colder. The water temperature was 52°F, the visibility was clear, and the Mighty Mississippi River was flowing south at a rate of 3 mph.

The S.S. Helena was steaming up the Mississippi River to Baton Rouge, Louisiana, her voyage having begun at the Panama Canal. At about 1130, near mile 91 AHP – Ahead Head of Passes, she took aboard Pilot Harold A. Rowbatham who was a member of the New Orleans & Baton Rouge Pilot Association.

Pilots are mariners with extensive knowledge of specific waterways. They are almost always contracted to come aboard a vessel to guide it through dangerous or congested areas. Once onboard, the pilot assumes navigational control of the vessel. Although the pilot is now guiding the vessel, the ship's master or captain maintains full responsibility over the vessel and can relieve the pilot at any time they feel the vessel is in danger.

It's noted that aboard the Helena, only the captain spoke and understood English beyond simple words and phrases. But that Harold Rowbatham had no difficulties in having his orders properly carried out by those on the bridge.

The Helena continued up the Mississippi River at various maneuvering speeds as required, but tried to maintain a maximum of 72 revolutions per minute on her propeller shaft, which gave the ship a speed of 14 miles per hour over the 3 mph current she was going against.

The pilot carried a portable transceiver tuned to 156.65 MHZ. This frequency is also known as "Channel 13" and is used for bridge-to-bridge communication between nearby vessels.

Vessels transiting the Mississippi River normally discuss meeting, passing, or crossing situations via Channel 13. If the other vessel cannot be reached via Channel 13, horn signals are given. One blast would indicate that I intend to leave you on my port side. Two short blasts would indicate that I intend to leave you on my starboard side. Upon hearing the one or two blast signal, the other vessel would, if in agreement, sound the same signal and take steps to affect a safe passing. If the proposed maneuver is unsafe, the danger signal of 5 or more short and rapid blasts would be sounded.

Using his radio Helena's pilot communicated with the M/V Betty Woods and tow at about 1230 and again at 1730 with the M/V Girlie Knight.

At 1705 the sun set and at about 1730 Helena's running lights were turned on and all but the stern and range lights visually sighted by both the helmsman and the lookout. The navigation light indicator panel showed all lights to be burning properly. This would appear to a vessel approaching Helena as a port & starboard red & green light with two white mast lights- one forward & one aft, indicating she was a power-driven vessel over 50 meters in length making way.

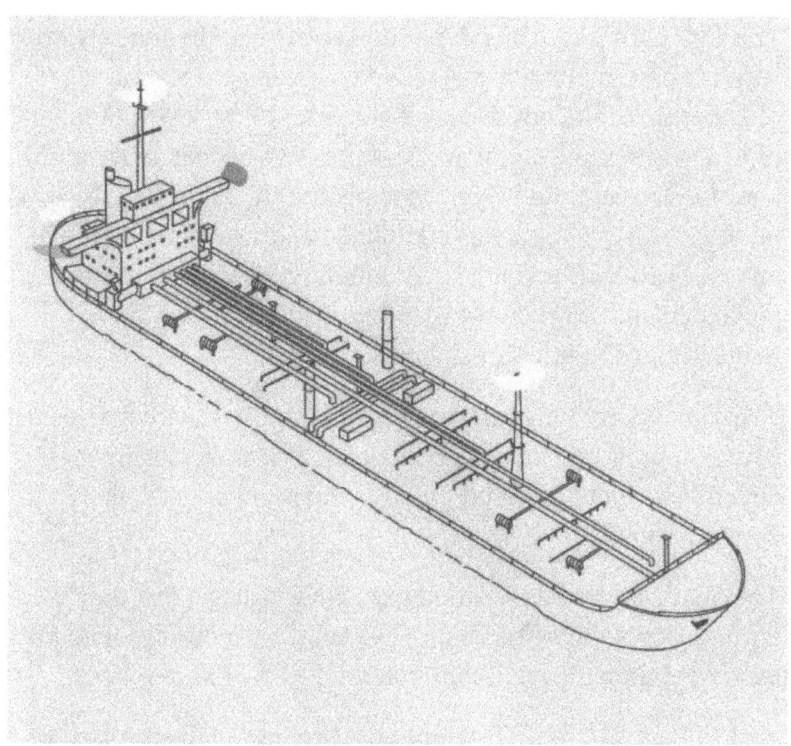

Lighting configuration for a vessel over 50 meters in length making way

She had two radars aboard. One was set on standby, and the other was inoperable. The one in standby was frequently checked by the mate of the watch.

At 1730 the bridge watch on Helena consisted of the pilot-Harold Rowbatham, a Third Mate - Wang Wei-Shue, a Helmsman – Wu Ko-Jen, and a Lookout – Chu Ching-Ling. Captain Shih Teh-Chang was in his cabin eating.

The C.G.C. White Alder departed her berth at the Inner Harbor Navigation Canal moorings in New Orleans at 1834 on the 6th of December. She got underway in order to retrieve a number of low water buoys in the Mississippi River, located between New Orleans and Red Eye Crossing at mile 224 AHP. By 1600 on the 7th this mission was completed and two 717 LR Buoys, one 620 LR buoy and nineteen unlighted buoys had been picked up and stowed aboard White Alder along with an undetermined number of sinkers.

These buoys are what marked the port and starboard side of the Mississippi Rivers navigable channel. The sinkers were concrete weights, weighing up to a thousand pounds, that held the buoys in place.

The White Alder had notified base New Orleans that they were enroute down the Mississippi River to their moorings with an estimated time of arrival of 0900 on the 8th of December.

At 1730 on the 7th of December the bridge watch consisted of CWO Samuel Brown and the helmsman, Seaman Roger Jacks.

The White Alder's standing orders stated that Channel 13 was to be guarded, meaning monitored, at all times while underway. There was also a regular procedure on board that Channel 13 was to be used to communicate with other vessels concerning meeting & passing situations.

Coast Guard regulations also required that the White Alder ensure that her radar was on at all times when underway in or near areas of reduced visibility and such other times as safety was deemed necessary. The White Alder had a standing order in place stating that the radar was to be on at all times while underway.

Standing orders also required White Alder to energize all navigation lights at dusk. This would have allowed a vessel approaching White Alder to see her port & starboard red & green lights along with a single white mast light. This would indicate that she was a power-driven vessel under 50 meters making way.

Lighting configuration for a vessel less than 50 meters in length making way

Another standing order required that the leeward pilot house door be open at all times underway. Divers reported that the

White Alder's starboard door, which would have been the leeward door, was open. This is to facilitate the hearing of another vessel's whistle.

At 1814 it was totally dark, and Helena was in the vicinity of mile 192 AHP, just below Bayou Goula Bend. This position of the river, between mile 192 and 197 AHP, turns approximately 180 degrees to the right for an upbound vessel and between mile 194 and 196 AHP, there is an island known as Bayou Goula Towhead.

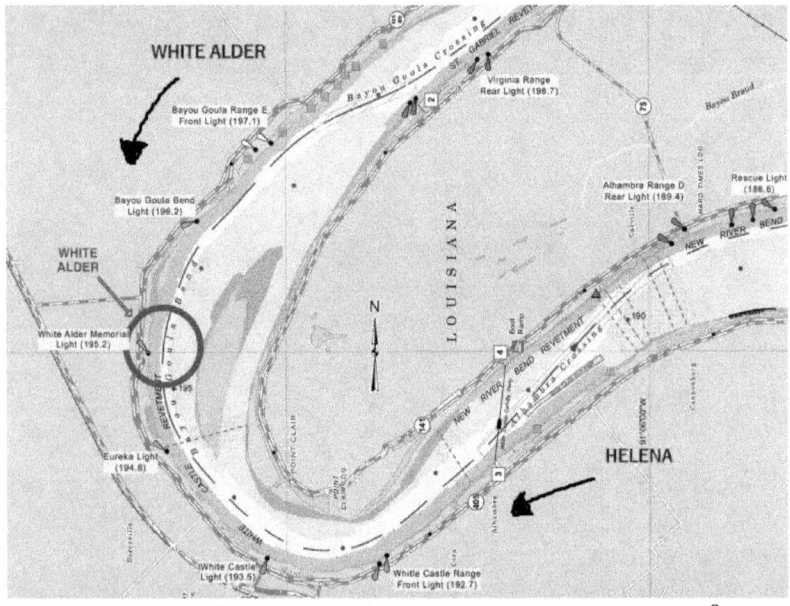

Bayou Goula Bend with collision location circled[8]

You can see Bayou Goula Bend in the previous picture with the direction of travel for Helena and White Alder. Note that each vessel needed to stay the right – starboard, side of their side of the channel.

As the White Alder approached the Bayou Goula Bend her bridge watch was as previously stated, CWO Brown and SN Jacks. The engine room watch consisted of EN3 Walton O'Quinn and EM2 Michael Agnew. BM2 Richard Kraus was in his rack – aka bed, located in the crew berthing area on the port side of the main deck amidships. FN Bruce Kopowski was sitting on the mess deck, located on the starboard side of the main deck. SA Lawrence Miller was heating up dishwashing water in the galley on the starboard after side of the main deck.

There is no indication as to how fast White Alder was proceeding. She did have a maximum speed of 9.2 knots, or 10.5 mph. Adding the 3-mph for the current would have bumped her max speed to 13.5 mph over the bottom. Divers did however find her rudder set at 10° right rudder and pilot house engine controls set for 1/3rd forward speed on both engines, which would be a little over 3 knots or almost 3.5 mph, plus the 3-mph current would make it 6.5 mph over the bottom. This would also be indicative of making the left, or port, turn downbound At Bayou Goula Bend and the expected speed she would have been going making such a turn.

The Helena was operating at her previously mentioned speed of 14 mph over the bottom. Her rudder was maintained through most of the bend to keep the vessel in an easy swing. At 1819 she was at mile 193 AHP.

At 1820 Helena was approaching White Caste Landing, about at the start of the upbound bend, and broadcast that she was upbound and approaching Bayou Goula Bend. This is a normal procedure in such an area to check for any downbound traffic and to alert such traffic to their presence. No reply was received. She had her starboard door and porthole windows closed, but her port (leeward side) door was open. As previously mentioned, this is done to hear for whistle signals.

As Helena proceeded with her turn, she favored the left bank, which is normal for Bayou Goula Bend due to a sand bar off the south end of Bayou Goula Towhead.

At 1822 the third mate & lookout both reported a red light with a white light over 15° on Helena's starboard bow, about 2 or 3 miles upriver. This would be the port side profile of the White Alder. Because White Alder was coming down bound and the turn veered to the right for Helena, this would be normal for a port to port passing.

Pilot Rowbatham on Helena sounded one long blast on the whistle to which no answer was received. He then tried to contact the White Alder with his portable transceiver on Channel 13, but there was no reply. These actions were in accordance with the 'Rules of the Road." Captain Shih Teh-Chang stated that he heard the whistle blast, but stayed in his cabin eating.[11]

As Helena swung right into her turn, the down bound red and white light of White Alder crossed the bow of Helena, opening to 10° to 15° on her port bow. Seeing the White Alder's port red running light, the pilot of Helena assumed there would be the normal port to port meeting situation. Helena's speed was maintained, and the White Alder was monitored.

With the exception of White Alder not replying to the Channel 13 call nor the whistle signal, everything looked normal.

At 1827, the Helena saw the green light and one white light of the White Alder. This would be her starboard profile.

This would indicate that White Alder was crossing in from of Helena. The following painting depicts how the White Alder would have appeared to Helena as she passed in front of her.

Painting by M. Blaset of S.S. Helena & C.G.C. White Alder

This change of light configuration was reported by Helena's third mate, who was manning the engine telegraph, to the pilot. Upon hearing the report, the pilot stepped out onto the port bridge wing to get a better view of White Alder, because she had been obscured by Helena's high bow. Once on the port bridge wing Helena's pilot heard the danger signal, 4 blasts on the whistle, from the White Alder. Hearing the whistle blasts Captain Shih Teh-Chang stood up in his cabin.

At 1828 Helena's pilot ordered all engines stop, and a few seconds later, at mile 195.6 AHP, Helena's bow struck White Alder on her starboard side about 2/3rds of the way aft. Crew members on Helena heard the sound of metal scrapping on both sides and saw objects passing down the side. The pilot stated that "he could see what appeared to be the bow section of White Alder protruding from the water."[10] Captain Shih teh-Chang stated he felt the collision and ran to the bridge.[12]

BM2 Richard Kraus said, "just before the collision I heard the captain, CWO Brown, yelled something, but I couldn't say what it was." He went on to say that he was in his rack, on the aft main deck, & heard 6 blasts on the White Alders whistle just before collision. The impact tore his rack from its fastenings and sent it hurling forward while knocking him to the deck. The White Alder took a heavy port list and the berthing area immediately flooded. He took a deep breath, and he was carried by the rushing water to the forward passageway. "I kept going up and up and finally I hit the surface of the water. I heard people talking and I feared that I would get hit by something. I suddenly saw a buoy and grabbed onto it."[11]

FN Kopowski was sitting in the mess deck, on the main deck, and heard the danger signal on the whistle and started to stand. Immediately after standing there was a collision and the bulkhead, aka wall, at the forward end of the mess deck collapsed aft, the lights went out, and the compartment flooded. He was knocked to the deck and struggled to get out of the compartment, clear of the White Alder, and to the surface. Once at the surface, he swam to a nearby buoy and was assisted by BM2 Kraus.

SA Lawrence Miller said he was heating dishwater in the galley on the starboard after side of the main deck. He heard 4 blasts from White Alders whistle and some men shouting, "get off the ship." He saw CS2 Morrison and EN1 Rollinson running from the mess deck toward the galley door leading out to the stern. The collision occurred and CS2 & EN1 fell to the deck near the door. He was able to stay on his feet by holding onto the sink. The lights went out and water rushed into the space knocking him around until he was completely disoriented. Finding an air pocket, he was able to fill his lungs and start swimming. He surfaced next to a ring buoy in the middle of the river. He grabbed the buoy, looked upriver and saw the White Alder's bow extending about 8' out of the water. The bow immediately sank, and he noticed a large ship further up the river. He made his way to BM2 Kraus and FN Kopowski.

Herman Pierre who lives next to the river said "I heard what I thought was an explosion. I was sitting in the kitchen at the time. I ran to the window, and I saw what I thought was a flare. That's all I heard; I saw a flash of light."

At 1830 Helena's engines were ordered too full astern

Helena's Captain, Shih Teh-Chang, stated that immediately after arriving on the bridge he took control of Helena from Pilot Rowbatham. Helena energized her search lights, which showed no sign of the White Alder, and Captain Shih Teh-Chang stated, "so I can do nothing." He also stated that no lifeboats were lowered because of the quick arrival of the Mature, the ship that took over rescue work. Helena was attempting to lower a lifeboat, but this was time consuming. Pilot Rowbatham contacted his office in Baton Rouge of the collision using his portable radio and requested that they advise the Coast Guard.

At 1837 Helena set anchor at about mile 195.7, approximately 300' from the area of the collision.

Once on the buoy, BM2 Kraus, FN Kopowski, and SA Miller stated they could hear the anchor chain of Helena running out as she anchored. BM2 Kraus went on to say that they were shouting for their shipmates, but none of them appeared.[13]

The buoy the three White Alder survivors were on drifted down river and grounded on the right descending bank of the river near mile 194.5 AHP.

The Army Corps of Engineers vessel Mature was assisting in the Corps Bank Stabilizing Revetment work at St. Gabriel some 4 miles above Bayou Goula when she heard word of the collision over the radio. Mature, commanded by Captain John Marlow, immediately headed downstream and was the first to arrive on scene. Once there, the Mature pulled aboard BM2 Kraus, FN Kopowski, and SA Miller at around 1910.

The survivors were then transferred to the ambulances aboard the White Castle Ferry Feliciana for further transfer to the New Castle Hospital. After initial treatment they were flown to the U.S. Public Health Service Hospital in New Orleans where they were admitted. BM2 Kraus & FN Kopowski suffered from minor bruises, abrasions and lacerations and were held for one day. SA Miller suffered from being scalded by hot water and was held for 10 days.

AFTERMATH

An extensive search for survivors continued the following day. Three large black canister buoys popped to the surface of the river in rapid succession, as thought they had been fired from a cannon, about 1:30 p.m. Sunday as they broke away from the sunken vessel.[14]

Iberville Sheriff Earl (Bo) Williams looking at debris from White Alder[10]

A search of the river turned up various articles from the White Alder including a wooden chest, pillows, a table, a television set and a large life raft with the name "White Alder" on the bottom.[15]

Helena weighed anchor and continued to Baton Rouge, Louisiana. While there, a damage survey was conducted by the Coast Guard and representatives of the owners of the vessel. The damage found in the area of the bow was minor and did not materially affect the seaworthiness of Helena.

The wreckage of the White Alder was located at about mile 195.2 AHP, lying athwart, aka side to side, the channel in approximately 90' of water with her bow towards the right descending bank. The Corps of Engineers survey vessel Lester F. Alexander marked the wreck with 2 buoys, one off the bow and one off the stern, and the C.G.C Wedge (CG-75307) and her barge (CG-68017) were positioned to commence diving & salvage operations.

They would support Coast Guard divers along with a Navy team of "Hard Hat" divers who were flying to Baton Rouge. They wouldn't be able to start diving until first light on Monday the 9th because two tons of equipment had to be transferred to a ship and brought to the wreckage site.[15]

Once dive operations commenced, on the 9th, they were unsuccessful due to the strong river current and cold water which kept the divers from reaching the bottom.

On the 10th the C.G.C. Salvia (WLB-400) relieved the Wedge, but she remained in the area along with her barge. Salvia determined with her sea scanner radar that the wreck was in one piece, but the divers were still unable to reach the wreck or the bottom.

Also on the 10th, the Coast Guard Board of Investigation hearing commenced in New Orleans consisting of 3 Coast Guard Captains and a Commander acting as recorder.

On the same day the Justice Department filed a $3 million suit against the owners of the Helena claiming that the vessel was not operated properly at the time of the collision and was caused by the negligence of the Helena. Subsequently the Helen's owners filed claim in U.S. District Court claiming the opposite, that the two vessels were proceeding to pass one another when the White Alder cut across the bow of Helena.

These claims seemed a bit premature since an investigation into what actually happened hadn't even begun.

The Coast Guard Board of Investigation started by interviewing the crew of the Helena. Captain Shih Teh-Chang, Third Mate Wang Wei-Sheu, and Seaman Wu Ko-Jen gave testimony that is stated in the previous chapter. With the exception that the Third Mate and Lookout stated that they actually never saw a physical ship, just navigation lights.

When lookout, Chu Ching-Ling testified on the 11th of December 1968 attempts to illustrate the collision with models were abandoned after objections that he had no training in navigation, did not know the river channel and could not judge ship position because the Helena was rounding a bend at the time.

Captain Shih Teh-Chang to the left testifying with his interpreter to the right[16]

Herold Rowbatham, the Pilot, appeared and answered some questions, but invoked his 5th amendment rights and did not testify further.

On the 11th of December a salvage contract was awarded to Merritt Division of Murphy Pacific Marine Salvage Company. They arrived on the 12th of December with personnel and equipment and began diving operations. They also found the diving conditions extremely difficult. But after attaching themselves to a 250-pound descending weight they were able to reach the wreck. This mode of operation severely limited the mobility of the divers, which was restricted to a radius of about

15', as this was the maximum length of the hose and lifeline they could carry, from the descending wire which was caught by a grapnel on the port bridge wing. The zero visibility required all work to be accomplished by feel and the strong current precluded any exploration or work away from the lee of the wreck. The divers determined that the vessel was in a nearly upright position and heading approximately 303° true, which placed the port side of White Alder down river.

FN Bruce Kopowski & BM2 Richard Kraus appearing before the Board of Investigation[17]

BM2 Richard Kraus & FN Bruce Kopowski testified after the crew of Helena and gave their testimony as it was also described in the previous chapter. The picture of them was taken as they were appearing before the Board of Investigation on the 13th of December, less than a week after the collision.

On the 14th the body of SN Roger Jacks was recovered from the pilothouse on the last dive of the day. And on the following day, the body of CWO Samuel Brown was also recovered from the pilothouse. Movement within the pilothouse was very difficult due to the large quantity of debris and wreckage throughout the space, but divers were able to determine that the pilothouse had some structural damage. The forward limit of this damage appeared to be approximately at the center of the starboard door, and it extended aft involving both the side and after bulkheads. The starboard after corner of the pilot house appeared to be set in and forward about 1'. It was further established by the divers that the pilot house rudder indicator showed approximately 10° right rudder and that the pilothouse engine control levers for both engines were in a position that would run the engines at approximately 1/3rd speed ahead.

On the 17th of December the wreck shifted, dropping into a hole caused by the scouring action of the strong current against the hull. This shift caused severe turbulence on the surface of the river and the buoys attached to White Alder to indicate her location disappeared.

On the 18th diving operations were resumed, but it was not until the 20th that the White Alder was again located in 80' of water. At this time, it was determined that the vessel had dropped approximately 34' below the river bottom and heading altered to approximately 220° true. The only part of the vessel found above the mud line was the upper 10' of the A-frame with approximately 70' of water over it.

On the 21st of December all diving operations ceased, and the Coast Guard did not plan to renew efforts to recover bodies. But pressure from the lost Coast Guardsmen's families and public officials forced the Coast Guard to try and lift the White Alder. They contracted with Bisso Marine Company of New Orleans to study the feasibility of lifting White Alder in order to recover more bodies.

Between the 30th of December 1968 and the 2nd of January 1969 attempts were made to sweep wires under the hull in order to raise and transfer the vessel to the bank. On the 30th a 70-ton barge mounted derrick was in position to try and lift the White Alder. After two dives, the divers found the White Alder and were able to get a line attached to it. But when strain was applied the line gave way. The divers were unable to work any longer due to the strong current.

The Coast Guard in their Board of Investigation Report stated, "These efforts were completely unsuccessful and were stopped." At this point all salvage attempts and efforts to recover the bodies of the White Alder crew were terminated.

On the 7th of January 1969 SA Lawrence Miller testified in front of the Board of Investigation. His testimony was as stated in the previous chapter. But he did add that there was an open porthole in the galley he was in, and he clearly heard the warning blasts from White Alders whistle.

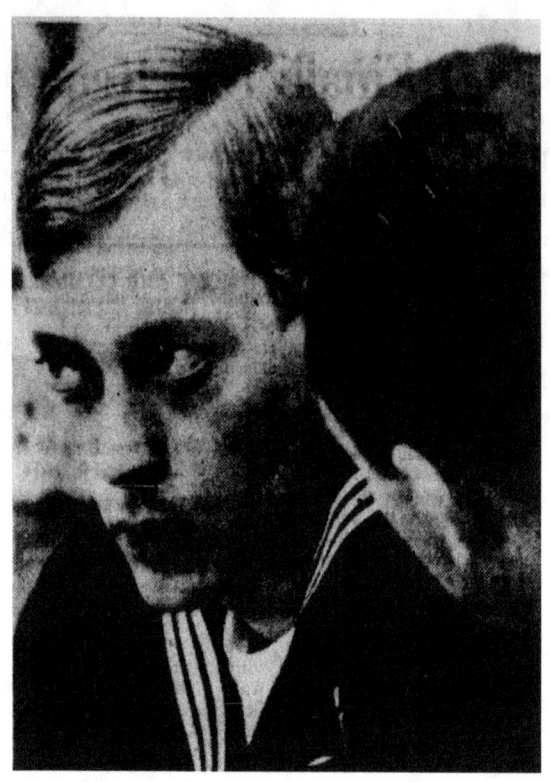

Seaman Apprentice Lawrence Miller testifying[19]

The Coast Guard had planned a Memorial Service on the 17th of January, but it was called off at the request of family members hoping their loved ones may still have their bodies recovered.

On the 5th of March a memorial service was held aboard C.G.C. Durable (WMEC-628) at the location of the White Alder. In attendance were the family of the White Alder crew along with the three survivors.

L to R - Pettey Officer Kopowski, SN Miller & BM2 Kraus aboard C.G.C. Durable placing a wreath over White Alder[20]

The body of Electrician Mate Second Class Michael R. Agnew was found in the Mississippi River on the 27th of July 1969. It was sighted near Burnside, about 10 miles south of White Castle, by a tugboat captain and recovered a short time later by Ascension Parish authorities. The body was taken to Donaldsonville where his remains were identified by his dental records along with a set of car keys and a knife found on the body.[21]

In June the White Alder's replacement arrived, the C.G.C. Zinnia WLM-255, A 122' inland Buoy Tender.

In September a privately sponsored fund, divided into 43 equal shares, each totaling $968, was distributed to the survivors of crewmen killed aboard the White Alder. The Coast Guard said trustees of the fund, totaling $41,654.00, were designated one share in the name of each of the 17 crewmen who died - one share for each widow, and one share for each of the surviving dependent children. The money was given to the widows of the married men and to the surviving parents of the single men.[22]

On the 7th of December 1969 the Coast Guard 8th District dedicated a park in memory of the collision. Rear Admiral Ross P. Bullard, commander of the 8th District, delivered the main address at the dedication of White Alder Park at the Coast Guard Base in New Orleans.

The Daily Advisor newspaper interviewed BM2 Kraus and SA Miller after the event, and here are their comments.[18]

BM2 Richard Kraus said he was working at Base St. Petersburg, Florida Aids to Navigation and will be getting out of the service in a month. He said "I'm looking forward to being out of the service. On this last assignment I requested this area but didn't particularly specify a job on land. But I'm glad I got it."

SN Lawrence Miller said "The only boat I have been back on was for the memorial ceremony in March. They didn't ask me if I wanted to go back on sea duty. They had an opening in the mailroom, and I took it." He was currently studying advertising at Tulane University and working part time at a flower shop. He still had 2-1/2 years left on a 4-year enlistment.

FN Bruce Kopowski, who had been promoted to Petty Officer, had already been discharged at this time.

In August of 1970 the National Transportation Safety Board met to compile its Marine Casualty Report. They evaluated reports completed by the Coast Guard and the Board of Investigation into a final report. The following is the analysis of that report:[1]

Determination of the actual cause of this collision is not possible due to the loss of all personnel on watch on the White Alder. Moreover, the declining of the pilot of the Helena to answer questions concerning the collision further limits availability evidence. The development of facts by the Marine Board of Investigation was also influenced by the decision that the U.S Government was not a party of interest. The Coast Guard did not designate itself as a party in interest. The investigation was also affected by the fact that the Coast Guard was investigating a case involving one of its own vessels. It is difficult for an investigation to be conducted adequately and effectively when one of the parties involved in a collision itself conducts the proceedings. This was demonstrated by the Marine Board's initial refusal to call witnesses who were familiar with standard operating procedures of the White Alder on the apparent ground that they were not involved at the time and place of the collision. This type of evidence of existing procedures is routinely used in accident investigations.

The Safety Board has noted in previous cases that the Coast Guard was investigating casualties in which the evaluation of important Coast Guard functions, other than operation of its own vessels, was involved. For example, in the foundering of the S.S. Panoceanic Faith, the search and rescue operations, communications, inspection, lifesaving equipment regulations, and lack of emergency signaling equipment were all Coast Guard responsibilities.

Although the bodies of CWO Brown & SN Jacks were recovered, no autopsies were performed to determine the cause of death and the physical condition of the men at the time of their deaths. Such a detailed examination might have revealed some physical incapacity which could have accounted for some of the unexplained actions which occurred in this casualty – specifically, the apparent lack of any response to radio calls, lights, or whistle signals, and the apparent lack of evasive maneuvers. This omission could not be corrected by the time the case was received from the Coast Guard.

The actions of watch personnel on the Helena are fairly well established by the records of investigation. However, the actions of the bridge personnel on the White Alder prior to the collision are only partially developed. Some facts can be determined from further analysis of the evidence which suggests several possible factors.

As Helena approached Bayou Goula Bend, the northeast wind tended to push her high bow to her left. This tendency was augmented by the effect of the current on her starboard bow. There is a sandbar extending south from Bayou Goula Towhead. Moreover, the normal trackline for upbound traffic favored the left ascending (west) bank. It is logical to conclude that she was favoring the left bank as she approached the bend at mile 194 AHP. At 1820, when she sighted the White Alder's red and white light, the Helena should have been visible to the bridge personnel on the White Alder. Helena's target aspect would show her green side light and white range lights, indicating she was heading left of the line of sight to the White Alder. The commanding officer of the White Alder might then have interpreted Helena's heading as indicating a starboard-to-starboard meeting. As Helena came right into the Bayou Goula Bend, her course changes should have been discernible by the closing of her range lights. The White Alder did not have range lights, hence her course changes were made more difficult to detect visually by personnel on the Helena.

The bridge personnel on the Helena were not concerned over the meeting, assuming a port to port passing. They proposed such a passing by sounding one long blast of the whistle. It was not until they saw the green side light of the White Alder that they recognized the danger of collision. The pilot was unable to raise the White Alder on Channel 13. Article 24 of the Western Rivers Rules requires the sounding of the danger signal, slowing or stopping until agreement for a safe passing is achieved. The pilot of the Helena maintained full speed until a few seconds before the collision. If he had slowed or backed down as soon as uncertainty developed, the collision would have been averted.

The actions taken by the White Alder personnel are difficult to analyze. Working back from the point of collision, which is established fairly accurately, for the White Alder to have been seen by the Helena personnel at 1822, the buoy tender must have been in the vicinity of mile 196 and favoring the west bank. The normal practice of CWO Brown was to favor the right side of the channel, under similar circumstances. The White Alder's speed from this probable position at 1822 until the collision at 1829 would have been about 4 mph plus the 3 mph current. This would correspond to the position of the engine order controls at one-third ahead. Also, a speed of 7 mph over ground would have enabled her to meet her 0900 estimated time of arrival at the Coast Guard Base, New

Orleans, the next morning. Apparently, neither CWO Brown nor Seaman Jacks heard the Helena's one blast signal at a distance of about 2 miles. If only the lee door of the White Alder's bridge was open, it is doubtful that the signal was audible, due to the White Alder being upwind and the noise level of her engines. It also appears that the bridge personnel did not hear the radio transmission on Channel 13, her VHF radio was inoperative, or Channel 13 was not being monitored.

Testimony of the White Alder survivors indicates that the only whistle signal initiated by the buoy tender was the danger signal just seconds before the collision. These witnesses did not notice any appreciable course or speed change prior to the collision. They did not hear any whistle signals from the Helena. It would appear that the White Alder could have taken evasive action. She was more maneuverable than the much larger Helena and had direct pilothouse controls and twin screws. If the one-third ahead speed and 10° right rudder conditions found by the divers were in effect at the time of collision, no drastic efforts were made to avoid the collision. This would lead to the conclusion that the commanding officer of the White Alder did not recognize the collision was imminent until too late to take corrective actions.

There is no evidence to explain the White Alder's crossing in front of the Helena. Several possibilities suggest hypothetical explanations. A steering failure or engine casualty might have caused her to cross the Helena's track. However, in the event of a steering casualty, it would seem logical to maneuver, using the twin screws to compensate. In the event of an engine failure, the rudder would be put hard over to compensate for the twisting effect of power on the only one shaft.

It is possible and likely that the captain of the White Alder did not realize that a collision was close at hand. He may not have seen the Helena until just before they collided. However, the running lights on the Helena were reported to be operating normally, and there were no known bright background lights along the bank to interfere with identifying the Helena's lights. There is a remote possibility that he confused the Helena's green starboard side light with the occulting green navigation light at Bayou Goula Bend. The White Alder operated only infrequently in this part of the river. It is also possible that the bridge personnel on the buoy tender were not adequately attentive, or that some distraction or incident interfered with maintaining an alert watch. However, CWO Brown was an experienced commanding officer, and would be expected to maintain an alert watch unless something abnormal occurred, particularly when approaching such a sharp turn in the river. It is possible that the White Alder's watch lost track of their position and started the turn too soon. The aids to navigation should have precluded this, but CWO Brown might have mistaken their identity. Another possibility was that a dizzy spell, seizure, or momentary blackout affected CWO Brown's actions just prior to the casualty. Nothing in his health records substantiates this hypothesis.

The fair wind and following current would also have affected White Alder's maneuvering the sharp turn left. To compensate for the set downstream, the buoy tender would have had to come left earlier than she would under no wind or current condition. The White Alder would tend to be set towards the west side of the bend, and probably experienced some crabbing effect in making her turn. This effect would be more noticeable at slow speed, as apparently was the case. The effect of the wind would be less on the White Alder than on Helena, due to the White Alder having less sail area above the waterline. However, there is a possibility that the wind and current caused both vessels to favor the west side of the bend. The commanding officer of the White Alder might not have anticipated his vessel's set, or the effect of the wind and current on the Helena. With both vessels set over on the west side, it is possible that the White Alder came left, as the commanding officer concluded that there was inadequate room for a port passage. He might not have taken this action early enough, or used enough speed or rudder to clear Helena.

None of these possible causes of the White Alder's actions can be substantiated by the record. However, the most probable of the several hypotheses appears to be that the commanding officer did not realize that collision was imminent until a few seconds before the casualty. No conclusion is apparent as to whether the Helena's movements were observed by bridge personnel on the White Alder. The next most plausible explanation is that CWO Brown misinterpreted the passing situation, and considered it to be a starboard to starboard passing. He could have assumed this if he did not hear the Helena's one whistle signal or VHF voice radio transmission.

There are a number of possible casual factors which resulted in this collision. A brief summary of these factors suggests measures to prevent recurrence of such a casualty.

Neither vessel took timely actions to avoid collision. Establishment of voice radio communications between the personnel on the bridges of the two vessels probably would have resulted in agreement for a safe passing. Proper evaluation of radar information available to personnel on both vessels would have shown the positions of these vessels with respect to which side of the channel the other vessel was favoring. If the White Alder had been fitted with an after range light, visual detection of her course changes would have been facilitated. The sharp bend in the river and relatively narrow channel at mile 195 AHP also contributed to the casualty.

The effect of the northeast wind and 3 mph current was a contributing factor. The prescribed rules for passing do not clearly provide track lines for upbound and downbound traffic. The system requires exchange of information by whistle signals and visual evaluation of the meeting situation. In this case, the system failed to achieve a safe passage, in spite of the fact that experienced personnel were at the conn of both vessels. Coupled with the whistle, a visual signal such as the amber light required by federal regulations might have alerted bridge personnel on the White Alder of the Helena's proposed one blast passing signal. Neither the White Alder nor the Helena was equipped or required to be fitted with an amber whistle light, under this rule, which is optional below mile 237 AHP on the Mississippi River. There was no positive way to know whether whistle signals have been heard by another vessel, unless a reply is received. The Safety Board has previously noted this problem in its action on the Marine Board of Investigation of the collision involving the SS Union Faith and M/V Warren J. Doucet, released on December 22, 1970. A recommendation was made in this report to the Coast Guard to consider specification of minimum performance standards for ship's whistles. The loss of 17 out of 20 crewmembers on board the White Alder was due to the buoy tender being overrun by the Helena, and rapid flooding of the smaller vessel. The Helena's shallow draft forward contributed to her riding up and over the White Alder. Scrape marks on both sides of the bow of the Helena indicate that she pushed the buoy tender under her bow in a very short distance and time. It appears that the angle of impact was between 60° and 90°, or more evidence of scraping would have been found on the side of the Helena. A smaller angle would have resulted in the White Alder glancing off on one side. The rapid flooding is verified by the testimony of the three survivors. It is probable that the other crew members were trapped inside the buoy tender. Except for CWO Brown and Seaman Jacks, they are still missing and presumed dead.

Author's note - The board apparently overlooked the body of Electrician Mates Second Class Michael R. Agnew being found in the Mississippi River on the 27th of July 1969.

Probable cause. The National Transportation Safety Board determines that the probable cause of this casualty was the White Alder's abrupt change of course across the bow of the Helena for unknown reasons. Also contributing to this collision were: the failure of the pilot of the Helena to sound the danger signal as soon as uncertainty developed concerning the other vessel's intentions: the failure of the officer in charge of the White Alder to sound a danger signal, followed by a proposed passing signal on the whistle, when the vessels were within one half of a mile of each other: the failure of the pilot of the Helena to slacken speed, stop, and reverse when risk of collision became apparent: and failure of the commanding officer of the White Alder to reduce speed, stop, and reverse prior to the collision. Other casual factors were: the failure of the White Alder to respond to the bridge-to-bridge radiotelephone communications initiated by the pilot of the Helena: the failure of the White Alder to respond to the Helena's proposed one blast passing whistle signal: the failure of both vessels' bridge personnel to make proper use of the available radar information: the sharp bend in the river at Bayou Goula Towhead: and the tendency for the current and wind to push the Helena's bow to her port in making the turn.

The Heavy loss of life on the White Alder was due to her being overrun by the Helena, and rapid sinking, trapping the Coast Guardsmen inside the hull.

That's the end of the report. And although it's very detailed, in my opinion it leaves a lot of unanswered question, which understandably the board agrees with and were lost when CWO Brown and SN Jacks died. I must admit that I do not understand why an autopsy was not performed on these two men. It would have ruled out a lot of speculation concerning a medical emergency happening. This may seem like a long shot, but I was on a Coast Guard Cutter and the executive officer, a lieutenant commander who was conning the bridge at the time, passed out with a major heart issue he never knew he had. We had to medivac him off the cutter via a helicopter.

Could CWO Brown have nodded off at a critical moment? White Alder got under way the previous evening at 1834. I have no record of who the qualified officers of the deck - OOD's, were. From my experience on a Mississippi River buoy tender this would have been the CWO & the BMC. But the BMC, Richard Batista, was left behind in New Orleans on leave. So, was CWO Brown solely conning the White Alder since the previous evening? I would assume he got some rest while the crew loaded buoys and sinkers during the day of the 7th, but what if he didn't.

Once again when I was on a Coast Guard River Tender, we had been working buoys all day. Just as we were getting ready to spud down – aka anchoring, for the evening the commanding officer, a CWO, received a call that a few buoys we had just set upriver had been taken out by a barge. The CWO decided to turn around and get back up to the buoys and spud down there. Then we would set them first thing in the morning and continue on our run-down bound. We hadn't gone far and all of a sudden, the buoy tender hit bottom and came to an abrupt stop. I looked out the starboard berthing area door and we were next to the riverbank. The CO had nodded off for a few minutes and we had run aground. It happens.

The absence of BMC Batista is never mentioned in any of the official reports I have read. Granted SN Jacks was in the pilothouse and it's unlikely they would have both nodded off. But SN Jacks had less than 2 years in the Coast Guard and would have been doing nothing more than taking commands from CWO Brown. And in his position, would not have been authorized to give or answer whistle signals from another vessel nor arrange passing information over Channel 13.

Not responding to the whistle nor the call on Channel 13 is also very concerning. I can understand not hearing the whistle call from Helena. White Alder was up wind and there was a 20-mph wind blowing. The White Alders pilothouse is also situated against the main diesel engine stacks. With both main diesel engines running along with the ship's service generator it would be hard to hear a whistle against 20 mph winds.

The call on Channel 13 going unanswered is very concerning. CWO Brown would not have been underway, especially at night, without an operational radio to monitor Channel 13. We know Helena's radio, or at least the one operated by Pilot Rowbatham, was working. He had just communicated with the M/V Girlie Knight at 1730. This would have presumably been for a passing situation and M/V Girlie Knight would presumably have been down bound ahead of White Alder.

Could CWO Brown have stepped out of the pilothouse missing the Channel 13 call? His cabin was located directly behind the pilothouse. He could have walked back to his stateroom to quickly use the head – aka bathroom. Telling SN Jacks to maintain the course and speed until he returned. I understand he probably wouldn't have casually gone to the head as he was approaching such a dangerous bend, but what if he had no choice and had to go use the head in his adjacent stateroom. When you look at the location of the pilothouse and the CWO's cabin on the ship's drawing profile, you can see this could easily have been done. I'm sure many a Coast Guardsmen on a smaller cutter like White Alder can attest to the OOD leaving the bridge to quickly use the head without a relief.

I would have to speculate that with the evidence at hand, CWO Brown had to make an emergency visit to the head. BMC Batista was not on board to relieve him, and he only had to go a few feet to his adjacent cabin. He told SN Jacks to maintain course and speed and he would be right back. He thought he would be finished and back in the pilothouse before the White Alder met another vessel and started into the Bayou Goula Bend. While he is gone, SN Jacks hears the call from Helena on Channel 13 and waits for CWO Brown to return to inform him. CWO Brown returns to find that White Alder is turning too soon into the Bayou Goula Bend and in imminent danger. All he has time to do is sound the emergency signal on the whistle. If he does shout out a command to SN Jacks to alter course and speed, he never has a chance to do so before the White Alder is struck.

I know this is all speculation, but I'm sure if you're a mariner reading this, you have a few speculations yourself.

CUYAHOGA

In 1978 there were two training centers producing commissioned officers in the United States Coast Guard. The Coast Guard Academy in New London, Connecticut was a 4-year college that graduated ensigns and Officer Candidate School -OCS, in Yorktown, Virginia which trained Coast Guard enlisted members in a 4-month course, graduating an ensign.

Each training center had a training vessel that was used to take trainees out for actual hands-on training in seamanship. The Academy had the C.G.C Eagle, which was a three-masted Barque. And OCS had the 125' diesel powered C.G.C Cuyahoga.

This phase of hands-on underway training is critical in the development of a Coast Guard officer. Because as an officer you are expected to, more than likely, qualify as a deck watch officer at your first assignment. And at some point, work your way up the ladder to command a vessel, referred to as a cutter in the Coast Guard.

Although there are other officer accession programs, such as ROTC and Direct Commission, the Academy and OCS are the more prominent. And as you read in the previous story about the White Alder, there is also the Chief Warrant Officer position, although they are not commissioned officers.

I myself obtained my commission through the Direct Commission program and attended a short 4-week course at the Academy in New London. I also had the opportunity to sail the C.G.C. Eagle in 1998 as a Senior Chief Petty officer on a cadet cruise. So, I can attest to the importance of getting underway as an officer trainee to learn valuable skills in seamanship and ship handling.

The incident you are about to read happened the 20th of October 1978 in the Chesapeake Bay between the C.G.C. Cuyahoga and the Argentinian flagged M/V Santa Cruz II. The aftermath was the sinking of Cuyahoga and the loss of 11 from her crew of 29. Santa Cruz, like Helana, was virtually unharmed.

As with the White Alder & Helena, there was a Marine Casualty Report[24] which I will outline in the following chapters. I will also lay out the events of the incident like in the previous story with newspapers reports and other supporting documents to help broaden the picture of what is going on.

This board was once again scrutinized because it was made up largely of Coast Guard Officer. As with the White Alder & Helena collision, there was lack of transparency and trust with the board being made up of 4 Coast Guard officers and 2 Department of Transportation representatives. As with the previous collision in this book, the Coast Guard was under the control of the Department of Transportation.

Since we have already gone over the "Rules of the Road" and navigational lights, let's get to know the two vessels that are involved in the fate of the Cuyahoga.

U.S.C.G.C. Cuyahoga (WIX-157)

C.G.C Cuyahoga was an Active-class patrol boat built in 1927 for the Coast Guard. In 1933 she was transferred to the U.S Navy and redesignated AG-26. In 1941 she was transferred back to the Coast Guard. She performed numerous duties for the Coast Guard until 1959 when she was assigned to the Reserve Training Center at Yorktown, Virginia as a platform to train officer candidates.

She was 125' in length, 23.5 feet beam, and was powered by two diesel engines powering two propeller shafts. She had radar aboard, which was located in the chart room, which was attached to the pilothouse.

Like the White Alder, she was commanded by a Chief Warrant Officer and had a Boatswain Mate Chief, Neal A. Verge, assigned as the executive officer. Her normal crew compliment was 11 permanently assigned members, who maintained the vessel. She also had an augmented crew of 4 who came aboard when underway. And she carried around 15-16, officer candidates who just got underway for training.

The crew of 29 aboard on the 20th of October 1978 was as follows:

PRIMARY CREW:

Chief Warrant Officer 4 (CWO4) Donald K. Robinson[27]

Chief Warrant Officer 4 (CWO4) Donald K. Robinson – Age 46 from Yorktown, Virginia he was a 20-year veteran of the Coast Guard and had been in command of Cuyahoga for a year. He had a lengthy career serving at sea as a Quartermaster sailing in the C.G.C. Yakutat, C.G.C Mackinac, C.G.C. 95314,

C.G.C. Yeaton, C.G.C 95304, C.G.C. Campbell, and C.G.C. 95332. As a quartermaster he would have spent his time on the bridge of these vessels plotting their course and navigating.

His previous assignment was aboard the C.G.C. Courier as a Chief Warrant Officer acting as the 1st lieutenant & deck watch officer. As the 1st lieutenant he would have been responsible for the deck division and as deck watch officer he would have been responsible for piloting the vessel.

Cuyahoga was his first assignment as the commanding officer of a vessel and on the 20th of October 1978 he had been in command of Cuyahoga for 16 months and 3 days.

While commanding Cuyahoga CWO Robinson had been reprimanded twice by the Coast Guard. First, after the Cuyahoga hit a drawbridge near Baltimore Harbor on May 31, 1978. And the second, when she hit a sea wall near the same harbor in April. For the drawbridge collision CWO Robinson was reprimanded for "poor judgement" and "poor seamanship." In the incident the Cuyahoga's radar antenna was knocked off the mast, causing $5k in damage.

He also received a letter of reprimand after the Cuyahoga struck a sea wall and dislodged a piece of granite.[29]

From my over 22-year career in the Coast Guard it is surprising that he wasn't relieved for lack of confidence on either one of those reprimands. But, because this was a training vessel, maybe he was given more leeway.

Machinery Technician Senior Chief (MKCS) David B. Makin – Age 34 from Newport News, Virginia. He enlisted in the Coast Guard in June of 1966, discharged in Jun of 1972, and reenlisted in June of 1977. He had a brother, Robert Makin, who was a retired Coast Guard officer.

Boatswain Mate First Class (BM1) Roger K. Wild – Aged 32 from Newport News, Virginia was the 1st lieutenant.

Substance Specialist First Class (SS1) Ernestino A. Balina – Age 35 from Newport News, Virginia. He enlisted in the Coast Guard in 1969 and was married to Leonie D.

Machinery Technician Second Class (MK2) Steven D. Baker – From Ashland, Maryland.

Quartermaster Second Class (QM2) Randy V. Rose – From High Point, North Carolina. He enlisted in the Coast Guard in 1975 and reported to Cuyahoga on the 2nd of October 1978.

Seaman (SN) Kevin J. Henderson – From Tampa, Florida.

Seaman Apprentice (SA) Jeffery T. Fox – From Conover, North Carolina.

Fireman (FN) James L. Hellyer – Age 20 from Newcastle, Pennsylvania. He enlisted in the Coast Guard in Aug 1977.

AUGMENTED CREW:

Yeoman First Class (YN1) William M. Carter[28]

Yeoman First Class (YN1) William M. Carter – Age 22 from Newport News, Virginia was qualified as an underway OOD. He enlisted in the Coast Guard in August 1974 and was the son of retired Army Lieutenant Colonel William C. Carter. He graduated High School from Denbigh High and was a Mate in the Sea Explorer Unit

Seaman Apprentice (SA) Michael E. Myers – Age 17 from Tacoma, Washington was a reservist on active duty for training.

Seaman Apprentice (SA) Michael A. Atkinson[30]

Seaman Apprentice (SA) Michael A. Atkinson – Age 18 from Spencerport, New York. He enlisted in the Coast Guard in March of 1977 and took his basic training at Cape May, New Jersey in August. Although listed as a reservist, he had been on Cuyahoga several weeks on active duty for training.

Seaman Apprentice (SA) David S. McDowell – Age 22 from North Chili, New York. He enlisted in the Coast Guard in March of 1978 and took his basic training at Cape May, New Jersey. From there he went to radioman school in California. He was a reservist on active duty for training while aboard Cuyahoga.

OFFICER CANIDATES:

Chief Warrant Officer Three (CWO3) Timothy C. Stone – From Grenada Hills, California was a 17-year veteran of the Coast Guard.

Lieutenant (Lt) Wiyono Sumalyo – Age 34 from Indonesia

Lieutenant (Lt) Jonathan E. Arisasmita - Was an exchange officer in the Indonesian Navy and a helicopter pilot.

Author's note – Sumalyo & Arisasmita are referred to as both captain & lieutenant in reports and newspapers. I am going to refer to them as lieutenant, which is probably more appropriate in a Coast Guard or Navy setting.

Aviation Technician First Class/Officer (AT1/OC) Candidate Earl W. Fairchild – Age 27 from Hialeah, Florida.

Aviation Technician First Class/Officer (AT1/OC) Candidate (AT1/OC) Robert P. Rutledge – From Tampa, Florida.

Aviation Technician First Class/Officer Candidate (AT1/OC) Lawrence V. Williams - From Alexandria, Virginia.

Machinery Technician First Class/Officer Candidate (MK1/OC) Edward J. Thomason – Age 32 from Wichita, Kansas. He enlisted in the U.S. Navy in February of 1966 and was discharged in January 1972. He enlisted in the Coast Guard in April 1977 and was single.

Radioman First Class/Officer Candidate (RM1/OC) Bruce E. Wood – Age 31 from Bellflower, California. He enlisted in the Coast Guard in July of 1965 and had a brother also serving in the Coast Guard, Petty Officer Third Class Jeffrey M. Wood, stationed at the Academy Hospital.

Aviation Technician Second Class/Officer Candidate (AT2/OC) Arne O. Denny – Age 26 from Portland, Oregon

Officer Candidate (OC) James W. Clark – Age 25 from Clovis, New Mexico. He enlisted in the Coast Guard in June of 1978.

Officer Candidate (OC) Peter S. Eident – From Falmouth, Maine

Officer Candidate (OC) John P. Heistand[25]

Officer Candidate (OC) John P. Heistand – Age 23 from Forked River, New Jersey. He had just joined the Coast Guard on the 10th of September after graduating from Northeastern University in Boston. He had a brother Lieutenant Commander Peter J. Heistand in the Coast Guard Reserve. And His father John, was a Coast Guard Academy graduate serving in WWII & Vietnam.

Officer Candidate (OC) Michael E. Moser – From Santa Barbara, California.

Officer Candidate (OC) Frederick J. Riemer – Marshfield, Maine.

Officer Candidate (OC) Joseph L. Robinson – From Carnegie, Pennsylvania.

Officer Candidate (OC) Earl G. Thomas IV

LEFT ASHORE ON LEAVE:

Boatswain Mate Chief (BMC)

Machinery Technician Second Class (MK2) James W. Blacketer III

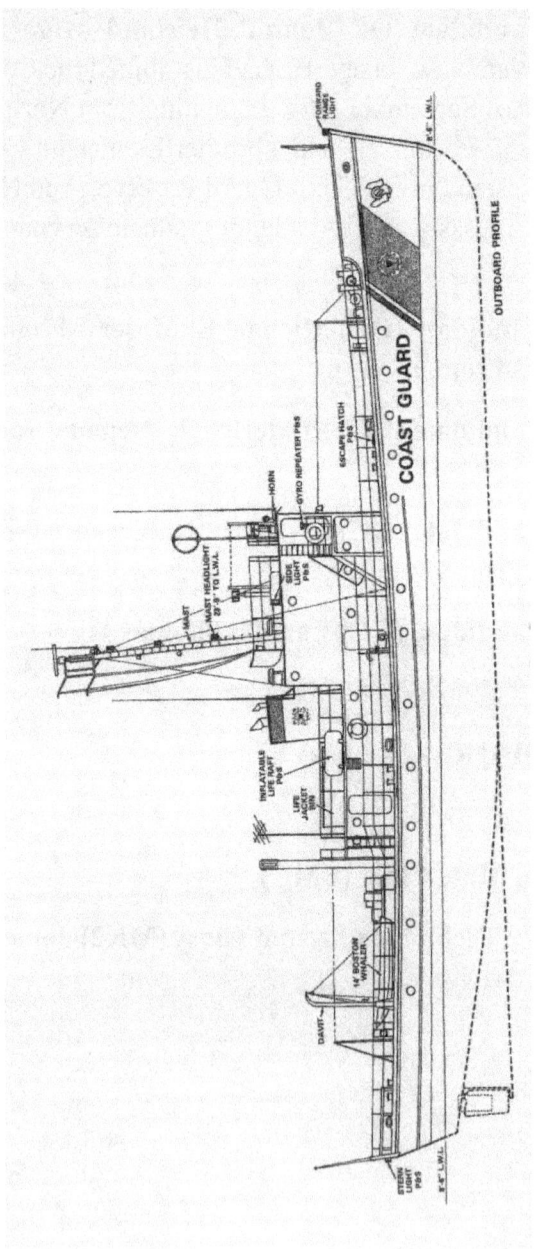

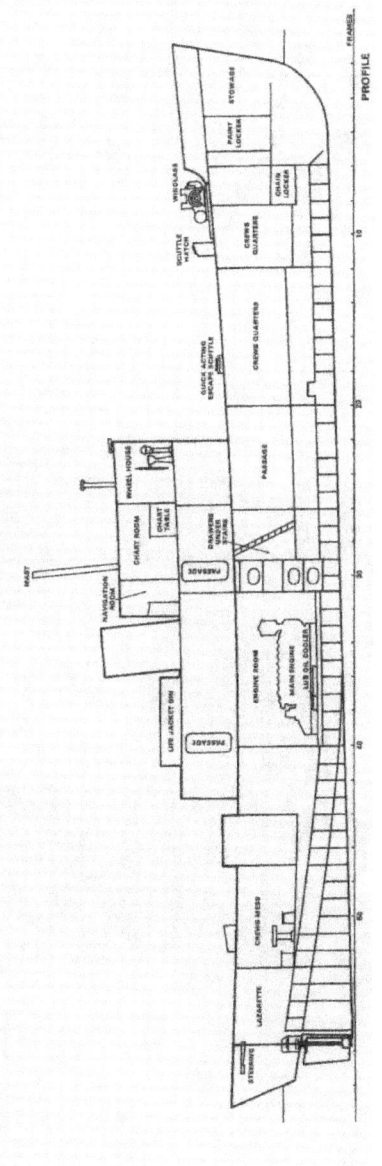

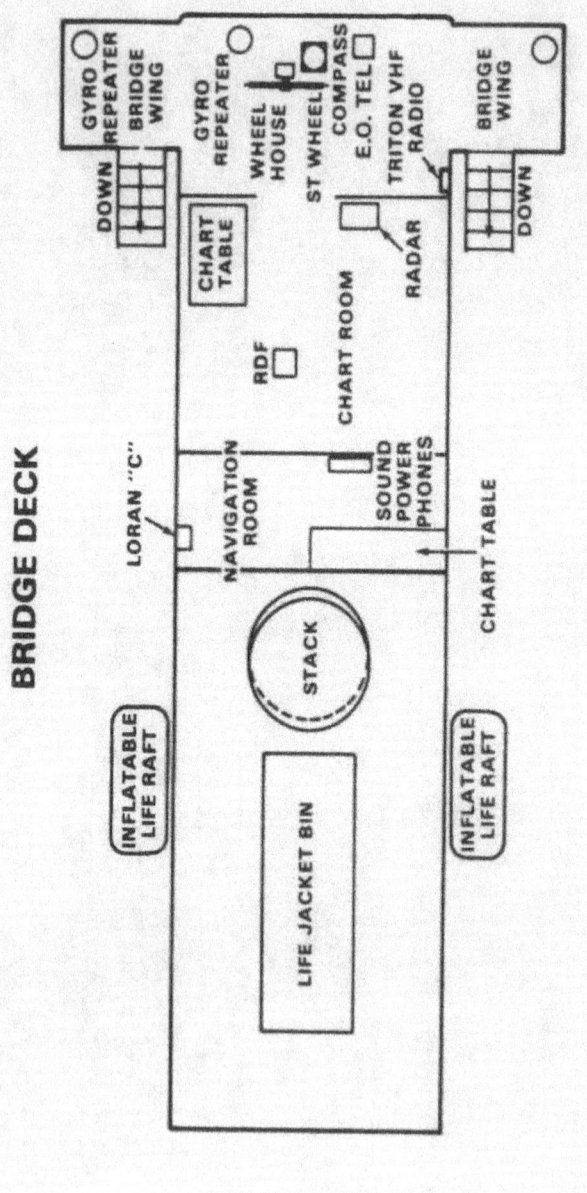

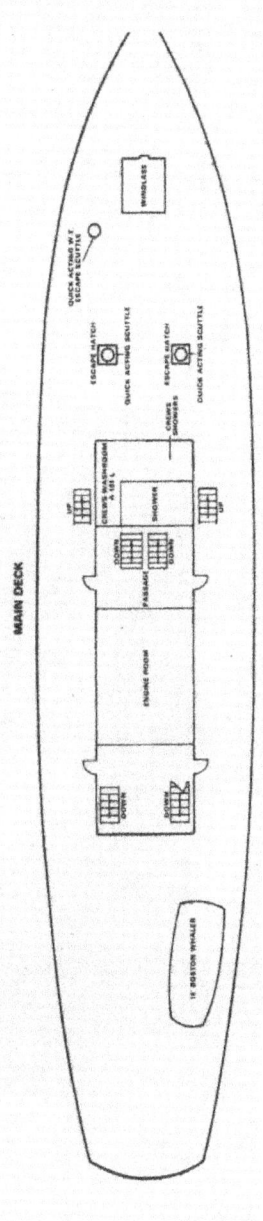

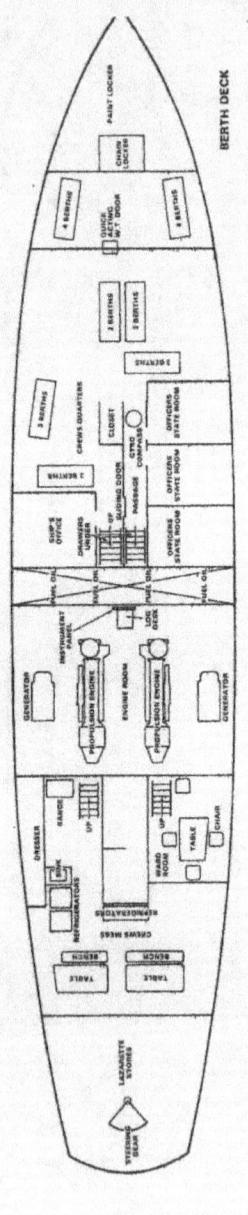

M/V Santa Cruz II[26]

M/V Sant Cruz II was an Argentinian flagged vessel, owned by the Argentinian government, built in 1977 and was brand new. She had a crew of 28, and at the time of the collision, presumed to be all Argentinian. She was 521' in length, had a breadth of 74', single propeller shaft, and was diesel powered.

She had a single pilothouse aft, located over the engine room. Control of the main engine could be accomplished automatically from the pilothouse command console, or from the engine room when in the manual or automatic mode., with orders passed from the bridge using a telegraph. The main engine is a large displacement, slow turning, Aesa-Sulzer six-cylinder diesel, capable of burning both heavy and light fuels, producing 9900 HP at 150 rpm. When the engine is being controlled from the pilothouse command console, changes in speed and direction are affected through a controller which regulates the rate of change so as to prevent damage to the engine. A stop order from a full ahead bell entered at the pilothouse command console will result in a progressive decrease of engine and propeller rpm's until the shaft stops turning about two minutes after the stop order.

Forward of the pilothouse are two sets of double cranes used in handling cargo hatches and 'tween deck pontoons. Two kingposts with a connecting crosspiece are located on the after part of the fo'c'sle; the forward range light is mounted at the center crosspiece. If you remember back in the previous story, Santa Cruz II would have the same nighttime underway lighting configuration as Helena with a forward and aft white range light – because she was greater than 50m in length and power driven.

Santa Cruz II was equipped with two independent radar units located in the pilothouse.

Like in the previous collision, the names of the crew who testified before the Board of Investigation were redacted. The following list was compiled from newspaper articles.

Captain - Abelardo Alboniz

Pilot - John Hamill

Chief Mate – Tomas Staino

Seaman – Christian A. Elliot-Grieve

Waiter – Oscor Pintos

THE 20TH OF OCTOBER

The Santa Cruz II had loaded up with 19,000 tons of coal and was getting ready to get underway from Curtis Creek, Maryland bound for Buenos Aries, Argentina. She had taken aboard John Hamill, a pilot with the Association of Maryland Pilots. As explained in the previous story, this was normal for vessels transiting congested areas that they were not familiar with.

Author's note – Curtis Creek is also the location of the Coast Guard's only Shipyard and a large logistics center.

The ship was late getting underway because a waiter aboard, Oscar Pintos, had fallen and cut his leg. He required medical treatment and once that was provided, Santa Cruz II got underway heading south out of Chesapeake Bay.

Pilot Hamill was an experienced pilot, and this trip was estimated to be around his 700th trip, guiding vessels in and out of Chesapeake Bay.

The Cuyahoga got underway about 6 times a year to sail on a 4-day cruise, training officer candidates. During the cruise the trainees would take on different jobs getting a feel for being underway, such as helmsman, radar operator, lookout, navigator, recorder, bearing taker, and even officer of the deck. Most of the officer candidates on this trip had probably never been at sea on a Coast Guard vessel.

Before getting underway on this trip the officer candidates received a pre-cruise briefing on Thursday the 19th of October. The briefing was conducted by BMC Verge and QM2 Rose. The briefing consisted of orientation of the Cuyahoga, watch quarter, and station bill assignments, discussion of watch position duties, use of equipment, and cruise itinerary.

The watch, quarter and station bill is a listing of assignments for casualties such as fire, flooding, collision and so on. It also details where you are to muster in the event there is a man overboard and which lifeboat you are to report to in the case of an abandon ship situation.

Also on the 19th, the day before getting underway, Cuyahoga's radar was repaired. It was having problems and was due to be replaced the day after getting back from this cruise.

As Santa Cruz II made her way south on the 20th, she maintained a speed of 13.4 knots through the water. On her bridge were Captain Albornoz, Chief Mate Staino, Pilot Hamill, a lookout on the starboard bridge wing and a helmsman at the steering console. Steering was set to the manual mode and the auto pilot was not being used.

Cuyahoga got underway from her homeport of Yorktown, Virginia at 1515 on the 20th. The vessel sounded the required long blast on her whistle, which worked properly. As she was making way it was noted that the gyrocompass repeaters were not working properly and were not following or tracking the true indication of the master gyro compass as the Cuyahoga's heading changed. The repeaters were aligned with the master gyro, and a comparison was made on a range as they transited the York River. The gyrocompass input to the radar, like that of the repeaters, was found to be unsynchronized and both were reset by QM2 Rose. After resetting them, the gyrocompass was found to have no error and functioning normally. The problem was presumed to be a faulty switch on the gyrocompass repeaters that was not actually clicking to the "on" position, and positive engagement of the switch needed to be verified when switching it to the "on" position.

As was normal procedure while transiting the York River, an abandon ship drill was conducted, and the crew donned their life preservers. Her planned itinerary was to proceed northward from York River in the Chesapeake Bay to anchorage in the Potomac River this evening. Tomorrow, Saturday, Cuyahoga would continue northward Baltimore, where it would moor overnight. Then return to the York River and anchor Sunday and return to Yorktown Monday.

At 1945 the bridge watch aboard Cuyahoga was as follows; CWO Robinson assumed the senior deck officer – SDO, SN Henderson assumed the quartermaster of the watch – QMOW, and QM2 Rose assumed the navigation supervisor. The officer candidates on bridge assumed watch training positions. They were Lt. Arisamita – port bearing taker, OC Thomas – radar operator, OC Eident, - navigator, AT1/OC Fairchild – OOD, OC Riemer – observer, OC Robison - helmsman, and AT1/OC Williams – navigator-recorder.

At 2000 it was dark and both Cuyahoga and Santa Cruz II had their navigational lights on.

Shortly after 2000 SA Myers was shaken from his rack and told he had to assume lookout watch above the bridge, which he did. He was joined by OC Moser, who was not on watch, but wanted to observe what was going on. It was SA Meyers first time to stand lookout. It was also noted that he had no training in this watch station.

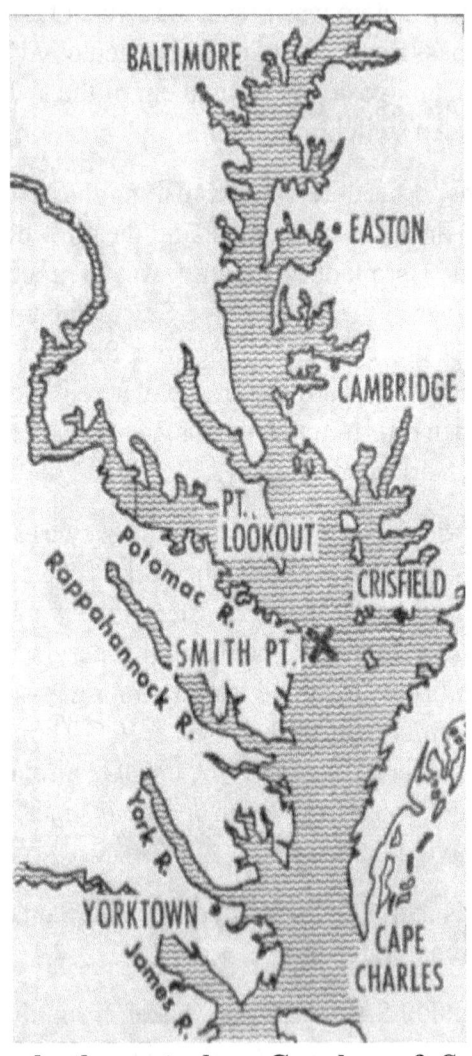

The "X" marks the spot where Cuyahoga & Santa Cruz II were at about 2100

At 2045 Cuyahoga was approximately 1.7 miles east of Smith Point Light, L.L No. 2725 on a course of 014°T and a speed of 11.8 knots.

As a note, *True bearing* -°T, is based on a circle of degrees and how a vessel relates to them, with True North as 000°, or 360, East as 090°, South as 180°, and West as 270°. *Relative Bearing*, as seen in the following diagram, is used for lookouts to identify objects in relation to the vessel.

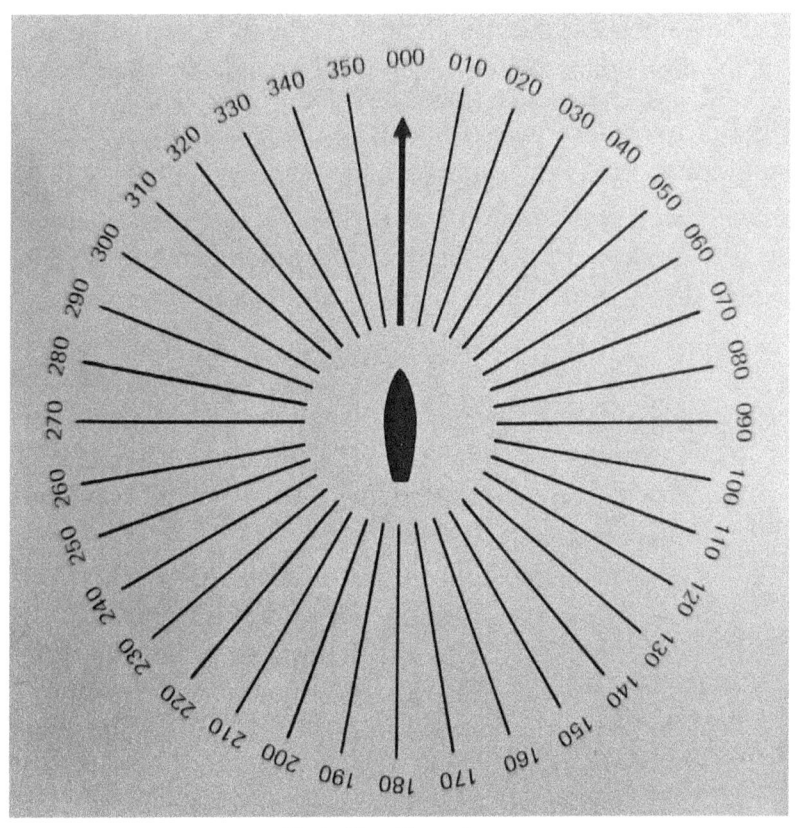

Relative Bearings

At 2047 Santa Cruz II passed abeam port to Lighted Bell Buoy 50, L.L.2726, at a distance of ¼ mile. Their speed had increased to 14.4 knots, the result of a following current. At this time Pilot Hamill changed course from 170°T to 163°T for the reach from buoy 50 to the Smith Point Fairway Lighted Buoy SP, L.L No. 2725.10, east of Smith Point. At this time Pilot Hamill had a look at the port radar and observed what he took to be a northbound vessel near the Fairway Buoy, as well as a southbound vessel near Smith Point Light. He observed a red side light, a white mast light, and other nondescript lights on the northbound vessel, which was Cuyahoga.

As I mentioned earlier, Cuyahoga would have the same light configuration as White Alder.

Light configuration Cuyahoga would be displaying

Pilot Hamill interpreted the red light of Cuyahoga to be a port to port passing situation and estimated the closest point of approach to be ½ mile. No whistle signal was sent nor received, and no call was sent or received via Channel 13. At this point Santa Cruz was 8 miles from Cuyahoga.

At about the same time OC Fairchild, training as the OOD, saw the lights, of what would be Santa Cruz II, and reported them to CWO Robinson. As mentioned previously, these would have looked identical to Helena in the previous story.

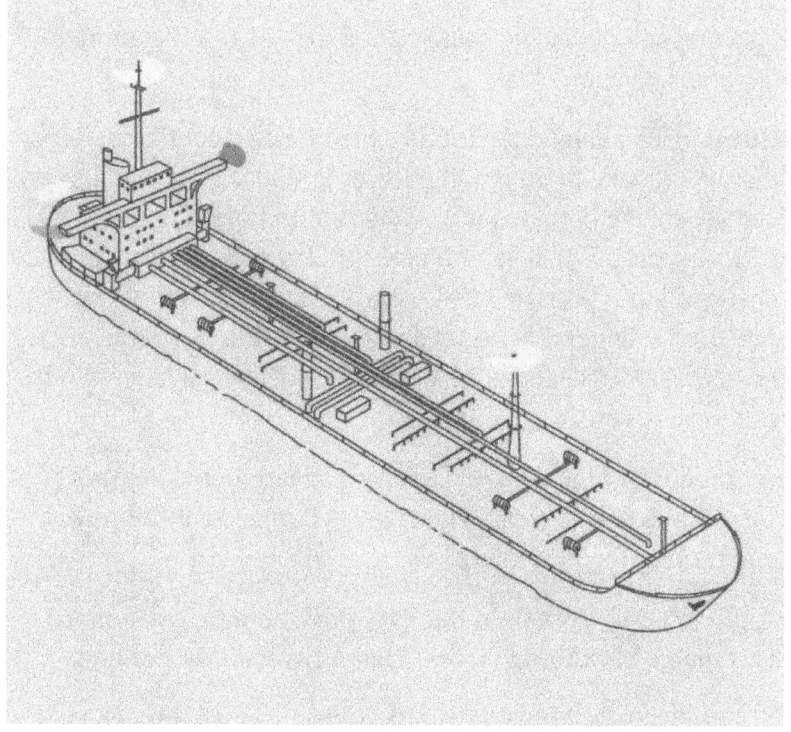

Light configuration Santa Cruz II would be displaying

Both CWO Robinson and CO Fairchild confirmed with binoculars that what they were seeing was in fact a vessel, and it displayed a red light and a white mast headlight. CWO Robinson then went to the chart room where the radar was located and determined that the vessel was 15,700 yards away.

Based on the small size of the radar contact and his perception of a single white light, he formed the opinion that the lights were of a small vessel proceeding in the same direction as Cuyahoga into the Potomac River. However, in reality this vessel was not traveling north in the same direction as Cuyahoga, but was the south bound Santa Cruz II coming towards Cuyahoga.

Cuyahoga's planned track line called for two course changes that would take them into the lower Potomac River, which you can see on the previous map. With Smith Point Light, L.L No. 2725, bearing 270°T at a range of 3900 yards, course would be changed left – port, to 338°T. With Smith Point Light bearing 199°T at a range of 5400 yards, course would be changed left – port, to 303°T. That course would then take Cuyahoga into the Potomac River.

At about 2049 the course was changed left to 338°T to allow for a tug & tow to clear. The engines remained at full power and turning for 11.8 knots.

At 2057 QM2 Rose used the VHF-FM radio to transmit Cuyahoga's location to Coast Guard Group Eastern Shore.

Shortley before 2100 Cuyahoga started to rotate officer candidate watch positions. AT1 Fairchild, the OOD, was to be relieved by OC Eident. The port bearing taker, Lt. Arisasmita, was to be relieved by OC Robinson. The radar operator, OC Thomas, was to be relieved by Lt. Arisasmita. The helmsman, OC Robison, was to be relieved by OC Riemer. The navigator, OC Eident, was to be relieved by OC Williams. The navigator-recorder, ATI Williams, was to be relieved by OC Thomas.

As you can see there were a lot of people on this small bridge. I would assume who was actually doing what was confusing.

When his turn came, OC Eident came forward from the navigation room and began preparing to relieve the OOD, AT1/OC Fairchild. As OC Eident's eyes adjusted to the dim light, AT1/OC Fairchild gave him the information necessary for him to assume the watch. AT1/OC Fairchild pointed out the lights thought to be the vessel ahead of them, which were actually Santa Cruz II. OC Eident looked at the lights through the binoculars and AT1/OC Fairchild advised him that CWO Robinson was aware of the vessel. OC Eident then recited the relieving information to CWO Robinson, who corrected him on a minor point and gave him permission to assume the OOD.

CWO Robinson then went out to the port bridge wing to verify the bearing of 199°T for their turn off of Smith Point. While there SA Myers, from his lookout position, made a report of a series of lights using the voice tube down to the bridge. OC Eident acknowledged the report, but being told that CWO Robinson already knew about the vessel, did not report it to him.

CWO Robinson returned to the bridge from the bridge wing and advised OC Eident to change course left – port, to 303°T. OC Eident relayed the command to the helmsman, and

Cuyahoga's course was changed. No whistle signal nor call on Channel 13 was given to the vessel in front of them.

At this time, 2104, Pilot Hamill on Santa Cruz II observed the lights of Cuyahoga change. He saw both the red and green sidelights at the same time - indicating Cuyahoga was headed directly towards Santa Cruz II, and then the green light alone – indicating her starboard profile. Cuyahoga's white masthead light continued to be visible throughout. He made comment of this to Chief Mate Staino and Captain Albornoz, who observed the same thing.

Pilot Hamill went to the port radar and switched scales from the 12 mile down to the 1-1/2-mile scale. Cuyahoga was seen as a contact at 330° relative, about midway in the scope and about 1 mile away. He placed the mechanical bearing cursor on the contact and went back to his position forward of the helmsman. Chief Mate Staino briefly looked at the radar and took up a position on the port side of the bridge by a window to observe. Captain Albornoz was at the starboard bridge window observing what was going on.

Meanwhile on Cuyahoga, CWO Robinson showed OC Eident where Cuyahoga was on the chart and directed him to take another chart back to the chart room, which he did. CWO Robinson then went out onto the starboard bridge wing.

At 2105, and about 1200 yards from Cuyahoga, Pilot Hamill on Santa Cruz II sounded one short blast on her whistle to signal Santa Cruz II's intent and that they would maintain course and speed as the privileged vessel. Their course of 163°T was maintained as well as their speed of 13.4 knots.

CWO Robinson on Cuyahoga heard the single blast from the vessel in front of him – Santa Cruz II. But still believing that the vessel was proceeding on a near parallel course that would take it into the Potomac River, CWO Robinson decided to alter course further to the left – port, to allow the other vessel to haul ahead on Cuyahoga's starboard side.

CWO Robinson said to OC Eident, "we are going to return that one blast, and I advise you to change course to 290°." OC Eident gave the necessary command, and the helmsman changed course to 290°T. OC Eident then proceeded aft to advise the QMOW. CWO Robinson returned one blast on Cuyahoga's whistle and returned to the starboard bridge wing. Cuyahoga's speed remained at 11.8 knots.

Santa Cruz II did not hear the whistle reply from Cuyahoga and at 2106, sounded a second whistle blast consisting of one blast.

Hearing the second blast from Santa Cruz II, CWO Robinson came into the bridge from the bridge wing and placed the engine order telegraph to the all stop position for both shafts and went back out on the starboard bridge wing.

MK2 Baker in the engine room answered the call for all stop on both engines and remained between the engine awaiting another possible call.

Immediately after sounding the second signal of one blast, Santa Cruz II sounded a danger signal of 5 short whistle blasts, immediately followed by a second 5 short whistle blasts and ordered the engine stopped and the rudder placed hard to left – port.

The heading of Santa Cruz II began to change to the left – port and her speed began to fall off.

The following picture shows Santa Cruz II and Cuyahoga as they approached each other in these past few minutes.

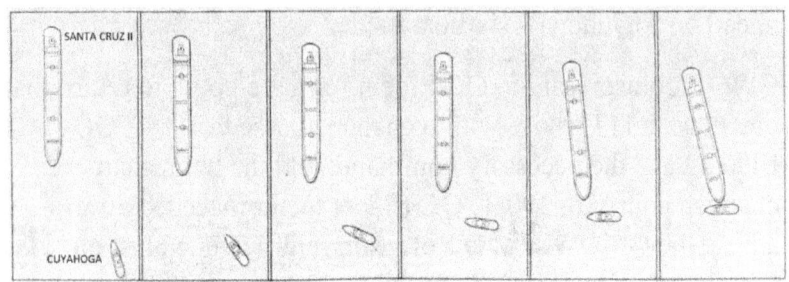

Santa Cruz II & Cuyahoga just before collision

You can see that Cuyahoga has been steering to the left – port, directly in front of the path of Santa Cruz II.

At 2106 the Cuyahoga heard the 5-whistle blast from Santa Cruz II. BM1 Wild was on his way up the starboard ladder to the bridge wing and shouted "Oh my God, Captain, he's going to hit us." Hearing this CWO Robinson placed the engine order telegraph at full astern for both shafts.

MK2 Baker, standing by in the engine room answered the call for full astern.

A few seconds later Santa Cruz II & Cuyahoga collided.

Cuyahoga heeled over to her port side 40 to 50 degrees, submerging her port main deck. Her hull, caught on Santa Cruz II was pushed back through the water at nearly 13 knots for 30-45 seconds as she slid down the starboard side of Santa Cruz II. This caused massive down flooding of the forward berthing spaces, engine room, galley, and after berthing areas. All the lights went out, except for 4 emergency lanterns in the engine room. Within 2 minutes Cuyahoga sank.

Santa Cruz II only sustained a penetration about 7 feet above the water line.

Drawing of Collision[34]

CWO3 Stone was standing on the ladder leading to the starboard bridge behind BM1 Wild when he saw Santa Cruz II about to hit Cuyahoga. He jumped from the ladder and ran to the rear port side of Cuyahoga to get as far away from the point of impact as possible. He held on to a stanchion and said a prayer "God give me strength." The force of the collision rolled Cuyahoga over on her port side and CWO3 Stone found himself under water, he had lost his grip on the stanchion. He quickly got his bearing and started to yell to shipmates to swim away from the Cuyahoga to avoid being sucked under with her.[31]

ATI/OC Earl Fairchild still on the bridge said "There was a loud rushing noise as we were dragged backward. The water started pouring in on the port side. It made a loud "whoosh" sound. Within seconds we were engulfed by white foam. Everybody was trying to hold onto something. I was hanging on by one hand to electrical cables. I saw a buddy disappear under the foam. There was wire wrapped around his leg. He was being dragged under by the ship. He let out a groan when he went under. My heart sank. I couldn't reach him. There were two others that were swept away, but I couldn't reach them."

He went on to say that once in the water CWO3 Stone was a hero as he yelled for people to swim to him at the Boston Whaler and kept them together as they waited for help.[32]

OC Earl Thomas was the navigator recorder and was making a notation and wearing a sound powered phone headset. As the Cuyahoga was hit and heeled over, he dived under the water. He found a door in the radar room, then another to the bridge. In the bridge he found a window, cranked it and swam to freedom.

SA Meyers and OC Moser above the bridge looked for a way to escape. OC Moser's first thought was to head for the life jacket storage, but Santa Cruz II's bow had blocked the way. Then he saw the 14' fiberglass Boston Whaler break free from Cuyahoga as she was going down. OC Moser swam for it. He reached it along with AT1/OC Rutledge. They removed the cover, finding it filled with water. The two started to bail it out with hollow rubber bumpers they found floating nearby.

SA Meyers said "It was chaos. We were trying to jump off, the ship was sinking fast and being crushed. I still can't believe I'm alive. The others who didn't get off never had a chance.[33]

SN Henderson was drawn beneath the surface by suction resulting from Cuyahoga going under and exhausted himself getting back to the surface. OC Robison came to his aid and assisted him to the Boston Whaler.

SA Fox said he yelled at FN Hellyer to hold on to something, but instead he ran to the stern of the ship. It was the last time he saw him.

Lt. Wiyona Sumalyo was sleeping in his berthing area.

AT2/OC Denny was also below decks asleep in his berthing area. He awakened to the shuddering of the Cuyahoga as she backed down in reverse. The lights went out, water rushed in. He scrambled to the ladder leading up to the main deck, but drawers between the ladder blocked his way. The emergency lights were not working, and water was now up to his waist. The next thing he knew he was standing on the starboard side of Cuyahoga in his underwear. He jumped in the water and found something to float.[34]

AT1/OC Rutledge was on the Mess Deck with MK1/OC Thomason watching TV. "The water rushed in over my feet and within a few seconds the compartment was flooded. I had time to draw a deep breath and swam over the object I was under. The lights were out, so I moved my hands though the water to stir up the phosphorous in the water to create a glow. I was able to make out the direction of the ladder, which was the only exit I knew. I could tell the ship was upright, and as I found the ladder, I felt around, but could not find MK1/OC Thomason and was running out of breath. Needing air badly, I swam along the ceiling and found an air pocket in the corner large enough for my head. I saw a hatch floating open and swam though it to the main deck. Moments later the Cuyahoga sank. I saw the Boston Whaler and swam to it."[35]

CWO Robinson also in the water was shouting for his men to get away from the Cuyahoga to avoid being sucked under with her. He went under several times but kept coming back up. He couldn't stay afloat on his own. Floating on his back didn't work, his head kept bobbing under. His arms were tired, and he called for help. Someone grabbed him and pulled him to a Styrofoam ice chest along with 4 others. From there they managed to paddle to the Boston Whaler.

When they reached the Boston Whaler, it was half filled with water. BM1 Wild had taken command and wouldn't let anyone in it except OC Moser and AT1/OC Rutledge who were bailing her out. After they finished, BM1 Wild ordered that CWO Robinson was to be put aboard her due to his serious health issues.

Once the Boston Whaler was stable, 5 men were put aboard while 8 clung to the side and 5 to the Styrofoam ice chest. They remained together awaiting help.

Following the impact Santa Cruz II continued to swing to port with the engines stopped. After Cuyahoga was clear of the starboard side, Captain Albornoz backed the engines half astern to take the way off the vessel. Pilot Hamill ordered the helm shifted to hard starboard, and set a course of 343º, to return Santa Cruz II to the point of collision, and the engine ordered to ½ ahead.

Captain Albornoz relieved Chief Mate Staino as officer of the watch and ordered the port motor lifeboat ready for lowering, and the accommodation ladder lowered. They then turned on the searchlight and located the Boston Whaler. They were able to launch a motor lifeboat with Chief Mate Staino and the Santa Cruz II's nurse aboard and it arrived at the Boston Whaler. The Cuyahoga men at the Boston Whaler told the motor lifeboat to continue on and look for other survivors.

The Santa Cruz II itself came alongside the Boston Whaler and a line was thrown to it. Unable to get the line close to the Boston Whaler, OC Moser swam to it and brought it back to the Boston Whaler. They were then able to pull themselves alongside Santa Cruz II.

Once they were close to Santa Cruz II, they were able to pull themselves up the freighter's accommodation ladder. When onboard the survivors received prompt and considerate attention. They were given dry clothing, blankets, food and beverages, as well as cigarettes and whisky taken from bonded stores. Those men with injuries were given first aid.

BM1 Wild said "I wasn't cold at all until we got on the Santa Cruz and held a muster. Then I realized how many people were missing."[36]

After looking for survivors, and finding none, it returned to Santa Cruz II only after Coast Guard helicopters and other assets took over search efforts.

Survivors aboard the Santa Cruz II were soon flown to Patuxent Naval Hospital for treatment.

Santa Cruz II remained on scene assisting in the search for survivors.

AFTERMATH

The following day, the 21st of October Coast Guard divers were sent down to ascertain if there were any survivors trapped inside Cuyahoga. They found her resting at the bottom in 58' of water on a heading of 225°T. She was upright and with a 25° list to port. The divers did not find any survivors.

Ltjg John Kercher, commander of the Port Huron, a Coast Guard 82' cutter assisting in the search, said there were no air pockets on the submerged vessel and the 11 missing men could not have survived the collision.

"There are no survivors, there are no survivors," he muttered to a throng of reporters on deck of the Port Huron. "The divers said there is no one alive inside the Cuyahoga."[37]

The survivors that were taken to Patuxent River Naval Station were released after they received treatment for minor injuries and exposure, except CWO Robinson. Although physically uninjured, he was kept for exposure and observation.

In the evening Santa Cruz II returned to Baltimore for repairs.

On the 22nd divers recovered the bodies of OC James W. Clark & FA James L Hellyer. OC Clark was discovered inside the main berthing area, near the starboard escape hatch at frame station 17. His death certificate would later state that he died of drowning. He was buried in Clovis Memorial Cemetery, Clovis New Mexico on the 28th.

FA Hellyer's body was found on the bottom, about 50' from the port bow of Cuyahoga. His death certificate would later state that he died from a fractured skull. He was cremated and his ashes were scattered at sea on the 7th of November.

On the same day Admiral John B. Hayes ordered the half-masting of flags at all Coast Guard facilities and on all ships.

On the 23rd the bodies of SA Michael A. Atkinson, OC Paul Heistand, SA David S McDowell and RM1/OC Bruce E. Wood were recovered.

SA Atkinson's body was found inside the main berthing compartment. His death certificate would later state that he died of a broken neck. He was cremated and his ashes were buried in Spencerport Cemetery in Spencerport, New York on the 27th.

OC Heistand's body was found in the main berthing compartment. His death certificate would later state that he died of drowning. He was buried in Veteran's Memorial Gardens at Hampton Virginia on the 26th.

SA McDowell's body was found in the athwartship passageway on the main deck, port side, by frame station 30. His death certificate would later state that he died from a broken neck. He was buried in Arlington National Cemetery on the 27th.

RM1/OC Wood's body was found on the port ladder leading up from the main berthing area to the athwartship passageway on the main deck by frame station 30. His death certificate would later state that he died from a fractured skull. He was buried in Acacia Memorial Cemetery, Woodinville, Washington on the 27th.

Also, on the 23rd, the Coast Guard Marine Board of Investigation commenced interviews at the Customs House in Baltimore, Maryland. Santa Cruz II crew were to testify first. The Board was made up of four Coast Guard officers. RADM R. H. Wood – chairman, Captain P. Nichiporuk – member, CDR J. L. Bailey – member, and LCDR J. E. Shkor – member & recorder

The first day of the investigation started off with the Santa Cruz II legal team objecting to the board being made up of 4 Coast Guard officers.

Kieron Quinn, the Baltimore lawyer representing Captain Alboniz said the Coast Guard would object if the board was composed of officials of the ship-owned company, Empressa Linas Marina Argentinas, which is owned by the Argentine government.

"I want to make it very clear that the Argentine ship owners have the same perception of this board" Mr. Quin said. Adding that a majority of the board should be commercial pilots. There is nothing, he said, to prevent the Coast Guard from hiring "private, independent marine experts" to conduct the investigation.

'The Coast Guard is investigating itself' said David R. Owens, the Baltimore lawyer for the pilot, Mr. Hamill.

Coast Guard Rear Admiral Raymond H. Wood, chairman of the hearing board, overruled their objections, saying "In our judgement, our authority is clear."

Capt. Abelardo Alboniz & his lawyer Kieron Quinn[39]

CWO Robinson refused to testify at the hearing. His lawyer recommended he not testify unless granted full immunity. He did, however, attend the hearings and shook the hand and embraced Sant Cruz II skipper Abelardo Albornoz and thanked him for helping rescue his crew after the collision. Abelardo said he knew Robinson would have done the same if the situation was reversed.

On the 24th, salvage operations were stopped due to 30 knot winds & 5-foot waves.

On the 25th, a memorial service for those lost on the Cuyahoga was held at the Reserve Training Center in Yorktown, Virginia.

BM1 Roger Wild (right) at Memorial Ceremony[40]

On the 26th the bodies of MK1/OC Edward J. Thomason, Lt. Wiyono Sumalyo, and SS1 Ernestino A. Balina were recovered from Cuyahoga.

MK1/OC Thomason's body was found just inside the engine room near the main deck athwartship passageway by frame station 40. His death certificate would later state that he died of drowning. He was buried in Dighton Memorial Cemetery, Dighton, Kansas on the 28th.

Lt. Sumalyo's body was found in the forward berthing area, compartment A-202-L. His death certificate would later state that he died of drowning. He was buried in Yogyakarta, Indonesia on the 1st of November.

SS1 Balina's body was found inside the boiler room, compartment C-201-E, frame station 42. His death certificate would later state that he died from drowning. He was buried in Peninsula Memorial Park, Newport News, Virginia on the 28th.

Also, on the 26th, the owners of the Santa Cruz II filed suit against the United States for $300,000 in damages to include $200,000 in repairs and $6,200 a day it is costing the owners to keep the ship, crew & cargo in Baltimore. The claim alleged that the persons in charge of the Cuyahoga were incompetent and inattentive to their duties.

This lawsuit was more than likely submitted because it was soon evident from testimony at the Board of Investigation that Cuyahoga had steered in front of Santa Cruz II.

QM2 Rose was a key witness and testified that when given the order to change course "It occurred to me to question the order, but there was too much action & too little time. What could I say? It wasn't a course I expected to turn to as far as a navigational course, but it was a firm order."[38] It was testimony like this that was starting to make it clear that Cuyahoga had in fact steered in front of Santa Cruz II.

QM2 Rose (left) & CWO Robinson (right) at the Board of Investigation[41]

On the 28th of October the body of YN1 William McDonald Carter was found floating on the surface by Coast Guard UTB 41358 about 500 yards north of the site where the Cuyahoga sank. His death certificate would later state that he died from a fractured skull. He was buried in Peninsula Memorial Park, Newport News, Virginia on the 31st.

On the 29th of October the body of MKCS David B. Makin was found floating on the surface by Coast Guard UTB 41330 two miles west of where the Cuyahoga sank. He was the last body recovered. His death certificate later stated that he died from a fractured skull. He was buried in Riverside Cemetery, Fairhaven, Massachusetts on the 1st of November.

I will have to admit that I was really surprised the autopsies for the majority of the dead crewmembers were from fractured skulls & broken necks – 6, verse the number of drownings – 5. I guess this is a clear indication of the tremendous impact of the collision.

C.G.C. Cuyahoga being raised by Navy Cranes

On the 30th of October Cuyahoga was raised by U.S. Navy floating cranes. It was a 7-hour recovery operation. Once out of the water she was placed on a barge destined for Hamton, Virginia. Once there Cuyahoga would undergo a thorough investigation. This was a pretty amazing accomplishment and enabled the detailing of findings which are to come.

On the 1st of November the families of SS1 Balina, MKCS Makin & OC Heistand filed a $1.9 million suit against the owners of the Santa Cruz II for the death of their servicemembers. They claimed the Santa Cruz II was traveling at an excessive speed and failed to post a bow lookout, sound the danger signal in at the proper time, put engines full astern, and use proper bridge communications prior to the collision.[42] Once again, this was more than likely initiated due to testimony at the Board of Investigation.

Although I could go on and quote testimony from the Board of Investigation, I think it is much better laid out in their assessment and findings report[24], which I will now quote, with my comments and paraphrasing as needed.

CUYAHOGA DAMAGE AND EQUIPMENT ASSESSMENT

The stem of Sant Cruz II struck Cuyahoga on the starboard side; the first contact was at the after starboard corner of the top of the wheelhouse by way of frame station 31, and then on the stack. The included angle between the centerlines of the vessels was about 45°. Initial contact at the main deck level was also by way of frame 31, which showed deformation of the deck coaming. The main deck was penetrated beginning at frame station 34, progressively deeper going aft to a depth of about three feet from frame stations 37 to 47. The side shell plating was punctured at the waterline by way of frame station 36, resulting in a hole one foot high by two feet wide, with a surrounding set-in. The area of set-in at and above the waterline continued aft to frame station 39 where the side shell plating was again penetrated. The penetration resulted in an opening that extended aft to frame station 45, and from the main deck down to a point four below the waterline.

The watertight bulkhead at frame station 40 separating the after accommodation and galley spaces from the engine room was breached, and set-in from the main deck to 2 feet below waterline.

The leading edge of the gross side shell penetration, from frame station 36 to 40 showed evidence of tearing and inward deformation. The side shell plating was compressed going aft and curled outward from frame station 40 aft to frame station 44.

Cuyahoga was equipped with a magnetic compass. The deviation table for the compass was out of date. There had been no adjustment of the magnetic compass or establishment of an accurate deviation table following the work done on Cuyahoga at the Coast Guard Yard in April 1978.

There were two ladders of wooden construction from the forward accommodation spaces leading up to the athwartship passageway on the main deck by way of frame station 30. The port ladder provided primary access to and from the ship's office, main berthing space, and forward eight-man berthing area. The starboard ladder provided primary access to the three staterooms. Both ladders contained drawers for storage built between the steps.

On impact, the drawers opened, making use of the ladders difficult, if not impossible. During the examination of Cuyahoga by the Board of Investigation as it lay on a barge after salvage, the drawers were found opened and firmly held in that position by swelling. Climbing the ladders was extremely difficult.

Testimony from Cuyahoga personnel was unanimous that the main deck watertight doors, port and starboard, at the athwartship passageways by way of frame stations 30 and 40, were material condition Yoke Modified fittings. These doors were routinely opened and closed by Cuyahoga personnel as the needs of comfort dictated. The natural and forced draft ventilation systems on the vessel were inadequate of themselves to maintain below deck habitability. In addition, excessive engine room heat encountered during warm weather operations would result in inverter fuse failures. There was no requirement to obtain permission to go through or leave open these doors.

Author's note - Coast Guard vessels adhere to a watertight and material readiness system of alphabetic fittings, doors, hatches, scuttles, portholes, ventilation systems, and so on. These alphabetic settings are for the most part to prevent the spread of fire & flooding. Underway, cutters are required to maintain at the minimum, material condition Yoke. These are all fittings classified "X" & "Y." These would include doors, hatches and scuttles. These fittings are not to be left open, unless logged open on the quarterdeck – bridge. Once passed through, they are to be secured.

Consistent with Coast Guard naval engineering practice and the compartment closure list, these doors were in fact, Dog Zebra fittings. As such, these fittings would be required to be closed only when condition Zebra was set, i.e., for battle and emergency conditions, or when darken ship was set. They would not be required to be closed for peacetime cruising when condition Yoke was set.

Testimony is conflicting as to whether or not the general alarm was working properly. MK2 Blacketer testified that it was not and had not been for some months; it was an item he intended to investigate and attempt repair. Other members of the crew testified that they believed the alarm to be working but could not remember recent use.

Author's note – Coast Guard vessels test all bells, alarms, & whistles at noon and prior to getting underway. Because Cuyahoga got underway after the noon hour, they should have been tested prior to getting underway and any discrepancies noted.

Cuyahoga was equipped with two ships service generators. However, there was no emergency generator or other central source of emergency electrical power that would automatically energize in the event of the loss of the ships service generators.

Cuyahoga was equipped with 4 relay lanterns; all were located in the engine room, two each per generator. There were no relay lanterns anywhere else on Cuyahoga.

Cuyahoga was equipped with portable type battle lanterns at other locations, including berthing and accommodation spaces and passageways.

The watertight subdivision of Cuyahoga below the main deck was compromised with various stuffing tube openings and other small penetrations in the after bulkhead of the engine room at frame station 40 separating the engine room from the mess deck and galley space and in the bulkhead at frame station 53 separating the mess deck and galley space from the lazarette. These openings were not sealed after removal of degaussing cables and miscellaneous wiring. Steering gear cables passed through openings in the after bulkheads under the main deck level. Such discrepancies were noted during the biennial inspections made by the Fifth Coast Guard District inspection staff in October 1975 and again in November 1977, but were not corrected.

There was not enough locker and drawer space for stowage of all personal effects of the officer candidates on Cuyahoga. A great deal of clothing and other personal effects of the officer candidates was hung from or laid on bunks or were otherwise loose. On impact, flooding, and submersion, this gear was strewn about the space.

During the inspection by the Board of Investigation of Cuyahoga salvage operations, it was apparent that this resulted in extreme difficulty during dewatering operations. The personal items, bedding, and other loose gear continually clogged pump strainers to the point that there were frequent and lengthy interruptions.

Evidence is conflicting with regard to the operation of the forward range light of Cuyahoga, which is mounted on the main deck bulwark at the stem. Testimony of Cuyahoga personnel is unanimous in that the light was on and functioning properly. Testimony from Santa Cruz II crewmen was that, as the vessel closed, a light was seen on the forward section of the ship; some thought it might be coming from a deck locker. There is no deck locker of the pilothouse on Cuyahoga.

Cuyahoga's port 15-man inflatable life raft was recovered by C.G.C. Point Huron at 0910, 22 October, 35 hours after the sinking, about 5.5 miles northeast of the collision. There were no people on board. The raft was fully inflated, and the drogue – like a sea anchor, was deployed. There was no evidence of damage or mechanical failure.

CUYAHOGA OPERATION & ADMINISTRATION

Watch training was conducted under the supervision of the senior officer of the deck, who was assisted in the pilothouse by a quartermaster of the watch. Navigation instruction was conducted by the navigation supervisor who performed his duties in the chart room and navigation room.

Underway watch training was conducted by assigning officer candidates to traditional watch positions under the supervision of assigned Cuyahoga personnel. The watch positions included an officer of the deck – OOD, helmsman, lookout, bearing taker, navigation recorder, radar observer, and navigator. If the number of officer candidates was insufficient, the watch was supplemented by assigned Cuyahoga personnel.

The degree of responsibility and authority of the officer candidate who was OC-OOD, varied as a weekend cruise progressed. Typically, an OC-OOD had no authority to maneuver the vessel during the Friday portion of the cruise. The OC-OOD would only make recommendations to the senior deck officer. This relationship would change; and on Saturday and Sunday the OC-OOD would be allowed more authority.

The underway navigational training on Cuyahoga on the afternoon and evening of the 20th was almost totally devoted to training in piloting. The bearing taker took visual bearings of aids to navigation using an alidade and gyrocompass repeater. The radar observer took radar ranges to prominent points of land and aids to navigation. The navigation recorder received information on sound-powered phones from the bearing taker and radar observer and he recorded the information in the bearing book. The navigator plotted the information received on a navigation chart to establish fixes at 5-minute intervals and made recommendations to the officer of the deck as necessary to keep the vessel on the inked intended track line.

The officer candidate navigation team had no assigned role with regard to acquiring, tracking, and plotting other vessels navigating in proximity to Cuyahoga. The efforts of the officer candidates were directed only at piloting. The officer candidates were given no instruction or requirement to plot radar contacts on the PPI scope or maneuvering board or develop closest-point-of-approach – CPA, information.

The navigation supervisor at the time of the collision had not been instructed that radar contacts should be noted and reported to the OC-OOD or SDO.

At the time of the collision, the navigation supervisor was Quartermaster Second Class Rose. QM2 Rose enlisted in the Coast Guard in September of 1975, served in the C.G.C. Reliance – a 210' medium endurance cutter, for 9-1/2 months as a seaman, attended quartermaster school, and was rated as a quartermaster in February of 1977. He served in Reliance in that capacity until transferred to Cuyahoga on the 2nd of October 1978. He was immediately assigned the position of navigator.

QM2 Rose testified generally about the mechanics of radar observation and maneuvering board solutions. He evidenced a general but not sophisticated understanding. His training in this area came from quartermaster school and from practical exposure in Reliance. QM2 Rose underwent no particular certification or testing prior to his designation as navigator on Cuyahoga. His proficiency in such matters was assumed, based on his advancement in the quartermaster rating.

Cuyahoga had undergone biennial inspection by the Fifth Coast Guard District inspection staff in 1973, 1975 and 1977.

The 1973 report, in commenting on the vessel personnel allowances of one officer and 10 enlisted, said in part:

The utilization of personnel is very good, but the workload of ship's maintenance, watchstanding and providing a training program is very taxing to the permanent party. To ensure the vessels' material condition does not deteriorate further additional personnel support must be furnished. This could be resolved by an increase in the personnel allowance or augmentation provided by the Reserve Training Center.

The 1975 report said in part:

The personnel allowance for this unit appears far below the necessary manpower needed to maintain this class of vessel, provide the routine administration, watchstands, and furnish qualified instructors for training OCS and Reserve personnel. Approximately two years ago, the unit was provided additional personnel support by augmentation of OC'S from the Reserve Training Center. This augmentation program was discontinued and only one extra billet is being furnished. This has placed a very heavy workload on the assigned personnel. This very reduced force also seriously limits the ship's force capabilities to provide adequate personnel to man underway stations during emergency and operational exercises when the ship's force is not augmented by trainees. During the underway training periods, the OCS and Reserve personnel are assigned Watch, Quarter and Station Bill billets and thereby furnish adequate manpower to meet the underway station requirement.

The unit is conducting a training program, although there is no documentation of this program other than a monthly training schedule of what subjects are to be covered for the month. The very low marks on the first aid and nuclear- chemical-biological warfare written examinations and the unsatisfactory grades in many drills reflect a need for improved training programs. It is recommended that the unit utilize the administrative volume of CCGD5 Training Program as a guide to develop an effective training program.

The 1977 report said in part:

The number of personnel allowed is not adequate for the maintenance, and for operations when no outside augmentation is available. Considering the Reserve training mission of the vessel, a yearly requirement for limited team training at Fleet Training Unit, Little Creek, is recommended. This will allow the permanent crew to remain proficient in the latest navigation, damage control, etc., techniques taught by the Navy. This training is at no additional cost to the Coast Guard.

Despite these reports, the personnel allowance list for Cuyahoga remained on the day of the collision the same as it was in 1973, 1975, and 1977. Cuyahoga had not undergone training at Fleet Training Unit, Little Creek, Virginia, nor was it scheduled to do so. Cuyahoga had not been visited by a

Coast Guard Ship Training Detachment, nor was it scheduled to be so visited.

The commanding officer of Reserve Training Center Yorktown and the officer in Charge of the Officer Candidate School both testified that they did not consider the officer candidates to be a supplementary resource enhancing the operational capability of Cuyahoga. In their view, the officer candidates were considered supernumerary trainees; the operational needs of the vessel with regard to navigation and piloting had to be met by the assigned personnel. As related by the commanding officer, Reserve Training Center Yorktown, the officer candidates were "sent aboard as observers and to the degree that their learning and skills and background permitted, to begin to practice some of the arts of seamanship."

Much of the time and attention of the senior officer of the deck was devoted to the underway training objectives for officer candidates of watch familiarization and navigation by piloting. The activities of CWO Robinson for 15 minutes before the collision are illustrative:

-on the port bridge wing working with bearing taker.

-listening to OC Eident make report prior to relieving OC/OOD of the watch and instructing with regard to proper relief procedures.

-on port bridge wing watching for turn bearing.

-advising OC Eident to change course left to 303°T

-discussing Cuyahoga's position and course to anchor on navigation chart with OC Eident.

-on starboard bridge wing; heard first signal whistle blast, sounded one blast and changed course to 290°T.

-stopped engines, heard second single blast and danger signals from Sant Cruz II, backed engines; Cuyahoga struck by Sant Cruz II.

The operational performance of missions by Cuyahoga was subject to routine oversight. However, there was no policy or program at Reserve Training Center Yorktown to have shipriders observe the underway operation of the vessel and training of officer candidates; the past practice of assigning an instructor from the officer candidate school to the cruise had been discontinued the preceding year.

Prior to 1974 both operational and administrative control were vested in commanding officer, Reserve Training Center Yorktown. By letters from the Commandant of the Coast Guard in 1974 limited administrative control was shifted to commander, Fifth Coast Guard District. The administrative control vested in the District Commander included all matters except personnel administration and pay. Commanding officer, Reserve Training Center Yorktown retained operational control of Cuyahoga, and control of personnel administration and pay.

The purpose and net effect of the change was to take advantage of the resources of the Fifth District. By virtue of its coterie of staff engineering specialists, it could provide more effective support, particularly with regard to maintenance.

CUYAHOGA COMMAND

CWO Robinson was ordered to assume command of Cuyahoga via orders from Commandant of the Coast Guard dated 30

March 1977, and he did so on the 17th of June 1977. Part of his orders were to obtain as much of the recommended training as possible prior to reporting for duty. Among the recommended courses for commanding officers were:

- Emergency shiphandling for senior officers

- Rules of the Road and Shiphandling (refresher)

Among the courses recommended for those who stand deck Officer Of the Deck watches were:

- Rules of the Nautical Road

- Emergency shiphandling.

CWO Robinson never received any of the recommended training because the instruction did not apply to vessels less than 143' or designated WIX.

However, CWO Robinson did receive a copy of the Command at Sea Orientation Manual CG-359, under a cover letter from Commandant Coast Guard, which states in part:

"This manual outlines the Commandant's policy concerning readiness training. If you have the opportunity, it is recommended that you obtain such training as you may find desirable before detachment from your present unit or while enroute to your new assignment."

CG-359 suggests:

-Rules of the Road

- Navigation and Shiphandling.

CWO Robinson testified that he had not undergone classroom courses or training in these matters. Rather, all his training was described as "practical" – on the job training.

CWO Robinson had the following sea service before assignment to the Cuyahoga:

U.S.C.G.C. Yakutat – 311', as a SA/QM3 for 33 months between 1952 to 1954.

U.S.C.G.C. Mackinac – 311', as a QM3/QM2 for 7 months in 1959.

U.S.C.G.C. 95314 – 95', as a QM2 for 12 months from 1959 to 1960.

U.S.C.G.C. Yeaton – 125', as a QM2/QM1 for 21 months from 1960 to 1962.

U.S.C.G.C. 95304 – 95', as a QMC for 27 months from 1965 to 1967.

U.S.C.G.C. Campbell – 327', as a QMC for one month in 1968.

U.S.C.G.C. 95332 – 95', as a QMC/QMCS for 19 months from 1968 to 1970.

U.S.C.G.C. Courier – 338', as a CWO2 for 24 months as a Deck Watch Officer from 1970 to 1972.

There are no records attesting to CWO Robinson's proficiency in seamanship, rules of the road, emergency shiphandling, or local knowledge.

CWO Robinson is myopic, with his most recent eye examinations showing 20/30 right eye, 20/100 left eye vision. His vision is correctable to 20/20 in each eye. He has sunglasses which were worn during daylight activity; these are corrective prescription eyeglasses. There is no record of his ever having obtained clear prescription eyeglasses for nighttime use. CWO Robinson was not wearing corrective prescription eyeglasses at the time of this collision.

The statement of CWO Robinson was to the effect that on initial sighting, he perceived a single white light and a single red light. However, Santa Cruz II was showing a white masthead light and white forward range light in addition to her red and green side lights. These lights were on and working properly.

There are no objective standards of visual acuity within the Coast Guard which reflect the particular demands of vessel command or deck watch officer duty. There is no procedure for identifying those individuals who must be wearing corrective prescription eyeglasses to meet eyesight requirements while performing such duties.

CWO Robinson was under a physician's care at the time of this casualty, being treated for possible allergic bronchitis and sinus problems. At the time of the collision the drugs Slophyllin, Alupent, and Entex were prescribed to him; a bottle containing Entex was recovered from his stateroom following salvage of the Cuyahoga.

Medical testimony received was that there was little likelihood that the drugs affected CWO Robinson's visual acuity, made him drowsy, or altered his personality. Absent CWO

Robinson's testimony, there is no evidence before the Board as to what medications may have been taken on the day of the collision.

Medical testimony further revealed that other than the Coast Guard aviation community, there are no distinctions made with regard to fit-for-duty status between shoreside personnel and those assigned to sea duty. Further, there is no distinction made with regard to those performing conning officer duties and other vessel personnel. The physician in this case candidly testified that he had never been on a Coast Guard cutter and did not have a sophisticated understanding of the duties of a commanding officer. However, his opinion was that there should be some differentiation for fit-for-duty determinations between people performing sea duty and those ashore.

COMMAND & CONNING OFFICER QUALIFICATIONS

Selection for command of a Coast Guard Cutter is based on appropriate prior experience. The evaluation is made by the officer of Personnel, which considers rank, career pattern, recency of sea service, and performance marks. A selection board annually selects prospective commanding officers in the grade of commander and captain. The primary criteria for selection is "performance record and appropriate prior experience." While the selection of warrant officers to command afloat is not the subject of board selection, the factors considered are generally the same. Command afloat for warrant officers is considered on the basis of:

-Availability of a two-year tour

-Previous experience afloat as a chief warrant officer

-Previous experience on the type of vessel to be assigned

-Local knowledge – desired but not required

-Individuals desire for assignment

-Good record

Author's note – It seems unusual to me that CWO Robinson would be given command afloat after being absent from sea duty for about 5-years. Although, his previous assignment was the 338' C.G.C. Courier, which was also homeported out of Yorktown, Virginia, so he would have known the operational area. And Courier was also used as a training vessel during the time CWO Robinson served on her.

Designation of an officer to serve as an underway officer of the deck on a Coast Guard cutter is a subjective determination made by the vessel's commanding officer. The factors considered vary from vessel to vessel and are subject only to the admonishment of Coast Guard Regulations, which state:

"….a person shall not be assigned duty as officer of the deck or as engineering officer of the watch unless in the opinion of the commanding officer he is qualified for such duty."

This procedure and criteria is distinguished from that required by the Coast Guard pilots and officers of the Merchant Marine. To serve as Master, pilot, or officer of the watch on a merchant vessel, the individual must possess a license issued by the Coast Guard. To obtain such a license, the pilot or merchant mariner must not only demonstrate appropriate prior experience, (i.e., service at sea) but must undergo written examination to demonstrate competence and professional knowledge, as set forth in Federal law.

Applicants for pilot licenses or endorsements are tested in the following subjects: Rules of the Road, Inland Rules, Local Knowledge, Chart Navigation, Aids to Navigation, Ship Handling, Chart Sketch of Route Waters, and Pollution Abatement.

Applicants for ocean master and mate licenses are tested in celestial navigation, sailing, lifesaving & firefighting, piloting, fuel conservation, international and inland rules of the road, and related subjects, as set out by Federal law.

BRIDGE-TO-BRIDGE RADIO

Testimony was heard from Captain Alexander Kaufman. He is an experienced master mariner, being the holder of a license to serve as master of vessels of unlimited tonnage, any ocean, and as a first-class pilot for most of the waters from Chesapeake Bay north to Narragansett Bay. He has been piloting in Chesapeake Bay since 1957, holding an unlimited Maryland pilot's license since 1963. He estimated that he had made at least 3,000 transits on Chesapeake Bay past Smith Point.

It is the practice of Maryland pilots transiting Chesapeake Bay on foreign merchant vessels to personally make all calls on Channel 13, VHF-FM. Typically, the pilot's portable radio is set to Channel 13 and the ship's radio is set to Channel 16. The reason for this is to minimize possible confusion resulting from language difficulties and misunderstanding of pilot intentions. The vessel's crew is not allowed to use the radio on Channel 13, VHF-FM.

The circumstances for use of bridge-to-bridge radio vary from pilot to pilot and are subjective. Pilot Kaufman, testifying as an expert, said that he saw nothing wrong with Pilot Hamill not using the radio while the vessels were some miles apart, because the red sidelight of Cuyahoga could be seen, and a clear port-to-port meeting was apparent. However, he testified that in smaller meeting situations he would consider use of the radio necessary if the closest point of approach would be ½ mile or less; he would desire the exchange of navigational information to be complete 10 minutes before the vessels reached their closest point of approach. He also testified that he had experienced a number of occasions when other vessels took sudden and unanticipated actions, changing proper and safe situations to hazardous situations.

COLLISION GEOMETRY

Based on the evidence adduced, the Board of Investigation plotted the tracks of Cuyahoga and Santa Cruz II as they navigated with respect to each other. Santa Cruz II sighted Cuyahoga when passing buoy 50 abeam to port. The range was nearly 8-miles, and Cuyahoga was within 5° of being dead ahead. Cuyahoga sighted Santa Cruz II shortly before coming left – port, to 338°T. The course change to 338°T put Santa Cruz II dead ahead at a range of 7.4 miles. The courses were withing 5° of being reciprocal, less than ½ point. Had not Cuyahoga changed course to 303°T some 15 minutes later, the vessels would have met port to port at a distance of less than 0.4 miles.

CONCLUSIONS

Shortly after Cuyahoga and Santa Cruz II came in sight of each other, they were in such relative positions as to establish a meeting situation. At a distance of 7.4 miles, Santa Cruz II was dead ahead of Cuyahoga; Cuyahoga was within 5° of being dead ahead of Santa Cruz II. The courses were within ½ point of being reciprocal, with a divergence of 5°. The vessels would have passed port to port at a distance of less than 0.4 miles had Cuyahoga not changed course. Both vessels had the duty to keep clear of each other and pass port-to-port. Pilot rule 80.3 would have required the vessels to sound whistle signals since they would have passed within half a mile of each other. It is the conclusion of the Board of Investigation that the rules pertaining to the meeting situation applied as Cuyahoga and Santa Cruz II continued to navigate with respect to each other.

Cuyahoga violated its duty to pass port-to-port, as required by Pilot rule 80.4, when, at 2104, it came left – port, from 338°T to course 303°T. This change put Cuyahoga on a course across the bow of Santa Cruz II and precipitated the collision.

The proximate cause of this casualty was the change of course left – port, to 303°T by Cuyahoga across the bow of Sant Cruz II. The improper navigation of Cuyahoga resulted from the failure of CWO Donald K. Robinson, Commanding Officer, Cuyahoga, to understand the position, true course, and speed of Santa Cruz II before changing course left - port, from 338°T to 303°T, which change put Cuyahoga on a collision course with Santa Cruz II. He failed to plot or require to be plotted radar contacts of vessels navigating with regard to Cuyahoga, either by grease pencil on the radar scope or on a maneuvering board.

CWO Robinson apparently perceived Santa Cruz II to be steaming in a westerly direction into the mouth of the Potomac River. This was a gross misinterpretation of the navigational lights of Santa Cruz II and the true situation, in which Santa Cruz II was steaming in a southerly direction. He persisted in this erroneous belief until the vessels were in the jaws of collision. This may have resulted in part from his erroneous belief that the small contact he observed on the radar must have been a small vessel, without regard to factors such as target angle or radar adjustment.

There is evidence that CWO Donald K. Robinson did on the 20th of October 1978, while serving in command of Cuyahoga, while underway in the Chesapeake Bay, negligently hazard said vessel by failing and neglecting to ascertain, or cause to be ascertained, the position, true course, and speed of Santa Cruz II, a vessel whose presence was well known to him, before changing the course of Cuyahoga left – port, from approximately 338°T to approximately 303°T, which change put Cuyahoga on a collision course with Santa Cruz II, as a result of which neglect Cuyahoga collided with Santa Cruz II and was sunk with multiple loss of life, all in violation of Article 110, Uniform Code of Military Justice.

Author's note – Violation of Article 110, Uniform Code of Military Justice is a serious offense. In the case of negligent hazard of a vessel, punishment may include dishonorable discharge from the Coast Guard and confinement at hard labor for two years.

Contributing to the cause of this casualty was the failure of CWO Robinson and Pilot Hamill to use bridge-to-bridge radio while Cuyahoga and Santa Cruz II were navigating with respect to each other. As Cuyahoga and Santa Cruz II closed for some 15 minutes, neither vessel initiated a radio call on Channel 13, VHF-FM.

CWO Robinson on Cuyahoga knew he would be taking his vessel across the shipping lane of Chesapeake Bay when he turned left – port, into the Potomac River. The added danger of this part of the voyage warranted special attention by CWO Robinson to the need to use the radio; no vessel would otherwise be privy to this information. CWO Robinson's neglect to broadcast his intention precluded Santa Cruz II from replying as to the danger of such a maneuver.

Contributing to the cause of this collision was the failure of Pilot Hamill to sound the danger signal when Cuyahoga's sidelights were seen to shift from red to red-green to green.

All during the time Cuyahoga closed Santa Cruz II on course 338°T, the intentions of Cuyahoga were properly felt to be understood on Santa Cruz II. It appeared that a port-to-port meeting would occur at about ½ mile; there was no reason to believe otherwise.

The change in course by Cuyahoga was known on Santa Cruz II, as the shift of Cuyahoga sidelights from red to red-green to green was clearly seen. The dangerous situation resulting from Cuyahoga's apparent departure from the rules and the port-to-port passage was evident to those on Sant Cruz II. The departure of Cuyahoga from the rules and its failure to continue to the anticipated port-to-port passage created doubt as to its intentions. Pilot Rule 80.1 provided that the danger signal shall be sounded immediately if, "when steam vessels are approaching each other, either fails to understand the course or intention of the other…"

After Cuyahoga changed course, Santa Cruz II was navigated as if the rules pertained to the crossing situation applied. Pilot Hamill improperly considered Santa Cruz II to be a privileged vessel obligated to hold course and speed and Santa Cruz II continued at full ahead on course 163°T. He sounded the two signal blast whistle signals to indicate his intention to hold course and speed. The danger signals were not sounded until nearly two minutes after Cuyahoga changed course, when the vessels were in the jaws of collision. It was only then that Pilot Hamill considered the vessels in extremis, and he stopped the engine and put the rudder hard to port.

Neither vessel was privileged during the initial meeting situation, and no privilege was created by Cuyahoga's departure from the rules. As much, Santa Cruz II was not privileged and was not obliged to hold course and speed. Rather, those in charge of her navigation were at liberty and had the duty to exercise their best judgement to avoid collision and act accordingly. The range of possible actions by Santa Cruz II included stopping, slowing, or turning, in addition to holding course and speed.

Had that judgement been exercised, no criticism could be made of the selected action; it is recognized that the mariner is not to be second-guessed in this regard. However, the failure to exercise judgement, which resulted from Hamill's improper belief as to privilege, was an error.

Had the danger signal been sounded by Santa Cruz II immediately after the lights of Cuyahoga showed her course change, it is possible that the collision could have been avoided. Cuyahoga would have been alerted that its actions were not understood and would have had an opportunity to take action. The single whistle signals later sounded by Santa Cruz II were inappropriate to the situation and did not convey the same urgency as would attend the danger signal.

There is evidence of violation of Pilot Rule 80.1 on the part of Pilot Hamill and Santa Cruz II with regard to delay in sounding the danger signal.

Contributing to the cause of this collision was the fact that Cuyahoga was not adequately manned to simultaneously train officer candidates and give adequate attention to the demands of safe navigation of the vessel. By virtue of the additional task of providing training and instruction, watchstanders were distracted from their navigation duties, or placed in inappropriate positions. The SDO, CWO Robinson, had to divert much of his attention from the needs of safe navigation. Instead of serving as quartermaster of the watch where he might assist the SDO, a position for which he was properly qualified, QM2 Rose was assigned to instruct officer candidates. SN Henderson was assigned as quartermaster of the watch, a position for which he had not yet been adequately trained.

As a result of the manning situation and the overall level of qualification of personnel in relation to their watch assignments, the safety of Cuyahoga was vested in the part-time attention of one man. Instructors for embarked trainees should have been provided to allow assigned personnel to attend the duties of safe navigation.

Contributing to the cause of this collision was the absence of a pilothouse radar display on Cuyahoga. A radar in the pilothouse would have lent itself to the monitoring of vessel contacts by the conning officer. With such an aid available, it is likely that CWO Robinson would have made better use of the radar and not persisted so long in his misunderstanding of the true position, course, and speed of Santa Cruz II.

Contributing to the loss of life in this collision was the fact that the main deck watertight doors on the port side by way of frame station 30 and 40 were open for ventilation. When Cuyahoga was pushed over by Santa Cruz II, these doors were partially submerged. As the two vessels continued through the water by virtue of the momentum of Santa Cruz II, a tremendous volume of water entered Cuyahoga through the doors. As a result of this rapid down flooding, personnel below were hampered in their efforts to escape against the rush of water. In addition, this condition resulted in Cuyahoga's rapid sinking, reducing the time in which those belowdecks could escape.

Contributing to the loss of life in this collision was the failure to sound the general alarm or use the public address system, which could have warned the personnel located below decks in Cuyahoga of the impending collision.

Contributing to the loss of life in this collision was the absence of adequate automatic emergency lighting in the passageways, berthing, and accommodation spaces in Cuyahoga, which absence inhibited escape from below decks in those minutes before Cuyahoga sank. It is probable that this lack of lighting contributed to the deaths of:

Lt. Wiyono Sumalyo

MK1/OC Edward Jerry Thomason

SS1 Ernestino Acogido Balina

OC James Wesley Cark

OC John Paul Heistand

Possibly contributing to the loss of life in this collision was the installation of drawers between the steps of the two wooden ladders leading from the forward accommodation spaces up to the athwartship passageway on the main deck by way of frame station 30. The open drawers may have impeded personnel trying to escape.

It is the conclusion of the Board of Investigation that the failure of the port life raft to deploy when Cuyahoga sank did not contribute to the loss of life in this collision. That is only the result of the fortuitous fact that the utility boat broke free of the sinking Cuyahoga and surfaced.

The Board of Investigation concludes that the port life raft did not inflate until sometime during the hours of darkness of the night of 21-22 October 1978. The Board of Investigation is unable to conclude which, among many possible factors, including insufficient depth, would account for the delay. Had the raft inflated promptly in this case, its utility to the survivors would have been diminished because of the absence of canopy lights and the fact that it would have drifted away. It is unknown why the raft did not remain attached to Cuyahoga by the sea painter.

In view of the technical nature of servicing life rafts and past experienced difficulties, it appears that the policy of frequent raft inspection by untrained personnel warrants reevaluation.

The Board of Investigation is unable to conclude whether any defect in CWO Robinson's professional knowledge and competence contributed to this collision. The absence of his testimony and lack of an objective system which records demonstration of professional knowledge and competence prevent this. Given the gross errors precipitating this casualty, the possibility cannot be discounted.

The lack of an objective system of record for vessel commanding officers to demonstrate their competence and professional knowledge is considered a weakness in the assignment process. It is recognized that present practice which identifies good past performance and seagoing experience is likely to result in satisfactory selections for command. It does not, however, serve to ascertain that selectees are, in all cases, possessed of desired levels of professional knowledge and competence at the time of assignment.

Similarly, the practice of deck watch officers being qualified only by virtue of observation on the vessel could be improved. While past results have been in the main satisfactory, there is lacking the degree of certainty that objective examination and periodic reexamination would provide.

The Board of Investigation is unable to conclude whether or not medications prescribed for CWO Robinson played any part in this collision. The possibility does exist.

The Board of Investigation is unable to conclude whether or not CWO Robinson's visual acuity and the absence of eyeglasses played a part in this casualty. Given the apparent initial perception by CWO Robinson of a single white masthead light and red sidelight on Santa Cruz II, and his impression that the lights were of a small vessel steaming westerly into the Potomac River, the possibility cannot be discounted.

The Board of Investigation considers the absence of specific standards of visual acuity for those performing deck watch officer duties or assigned as vessel commanding officers to be a weakness in the system of selection and qualification of such individuals. The Board of Investigation considers the absence of a system of record identifying those individuals who should be wearing corrective prescription eyeglasses while performing deck watch officer duties to be a similar weakness.

The efforts of CWO3 Stone, BM1 Wild, QM2 Rose, OC Moser, OC Robison, and the officers and men of M/V Santa Cruz II, with regard to post-collision survival and rescue are commendatory and worthy of note.

The deficiencies identified with regard to material and manning conditions on Cuyahoga are, to a large degree, the cumulative result of past management decisions, no one of which could be labeled as specifically causing this collision. The analysis of such decisions is beyond the competence and charter of this Board. Commanding officers at all levels are confronted with the need to accomplish assigned missions in an environment where needs outstrip resources. They must judge the impact of deficiencies and shortfalls on unit ability to accomplish broad missions and specific tasks. They must allocate resources among competing priorities. A significant factor in these management decisions is the adaptable nature of the ship and absence of objective standards of seaworthiness in the broad sense. Other than the hull, propulsion, steering gear and a cadre of essential personnel, there is little on a Coast Guard cutter that can be labeled vital. A vessel can sail on a specific assignment despite a minor reduction in capability occasioned by the absence of a particular man or the inoperability of a particular piece of equipment. Yet, the cumulative effect of several such reductions is a substantive denigration of seaworthiness.

Only after a tragedy such as this is an analysis of those factors which constitute seaworthiness in a broad sense undertaken. One such analysis into safety of afloat training for entry-level personnel in the Coast Guard has recently been concluded. To the degree that this Board of Investigation results in the establishment of objective standards of operational seaworthiness on Coast Guard vessels and the allocation of resources necessary to achieve those standards, the tragic loss of life in this casualty will not have been in vain.

Except as noted above, there is no evidence of actionable misconduct, inattention to duty, negligence, or willful violation of law or regulation on the part of any licensed of certificated persons; nor evidence of failure of inspected material or equipment; nor evidence that any personnel of the Coast Guard or any other government agency or any other person contributed to the collision.

This concludes the report by the Coast Guard Marine Board of Investigation. They did go on to make recommendations concerning CWO Robinson, Pilot Hamill, Coast Guard & Federal regulations, and several Cuyahoga crewmembers.

CWO Robinson was brought to Court Martial and was subsequently charged with the much greater offense of manslaughter, destruction of government property, and neglect. The first 2 charges could have led to up to 5 years in prison. In June of 1979 the manslaughter and destruction of government property were dropped due to legal issues that the Coast Guard declined to pursue.

The trial started in late October 1979 and lasted 2 weeks. CWO Robinson was subsequently only convicted of dereliction of duty, a lesser charge of neglect. The jury of 7 Coast Guard commissioned officers could have sentenced him to as much as 3 months in prison at hard labor and dismissal from the service, which would have cost him his military pension and benefits. But his punishment was a letter of reprimand and a token reduction in seniority, which saved his military pension and benefits. He retired soon after.

Pilot Hamill was recommended to be investigated under civil penalty procedures.

CWO3 Stone, BM1 Wild, QM2 Rose, OC Robison, and OC Moser were recommended to be considered for recognition for their actions following the loss of Cuyahoga.

The Board of Investigation recommendations concluded with several recommendations to change Coast Guard and Federal regulations governing Rules of the Road and watch standing.

In late December of 1979 District Court Judge Stanley Blair found CWO Robinson "solely to blame for multiple errors of judgement and perception, compounded by his absolute misapplication of each of the Rules of the Road." Judge Blair ruled that the $350,000 civil suit brought on by the Santa Cruz II owners against the Coast Guard was actually the fault of CWO Robinson, and not the Coast Guard itself, effectively denying Santa Cruz II owners a substantial monetary award.[43]

And in February of 1980 the Coast Guard dropped all charges against Pilot Hamill. The Commandant of the Coast Guard, Admiral John B. Hayes said, "In light of the decision of the District Court in the civil litigation arising out of the Cuyahoga/Santa Cruz II collision, I have determined in the interest of justice, to not proceed against Pilot Hamill."[44]

As for the fate of Cuyahoga herself. On the 19th of March 1979 she was towed 30 miles east of Cape Charles, Virginia where she was sunk as part of a Naval training exercise.

BLACKTHORN

A little over a year after the Cuyahoga collision the Coast Guard experienced yet another collision involving one of its cutters and a merchant vessel. While transiting Tampa Bay, Florida C.G.C. Blackthorn collided with the 605' tanker S.S. Capricorn. The aftermath was the total loss of the Blackthorn and 23 of her 50-man crew.

Once again there was a Board of Investigation conducted by the Coast Guard, along with a Marine Casualty Report[45], which I will outline in the following chapters. As with the previous two collisions, I will lay out the events of the incident and I will add newspapers reports and other supporting documents to help broaden the picture of what is going on.

The U.S.C.G.C. Blackthorn (WLB-391) was a 180' seagoing buoy tender which accomplished aids to navigation tasks similar to those of C.G.C. White Alder. Like the White Alder, she was considered a black hull, due to her hull being painted black. Coast Guard cutters are painted either white or black, except for the two or three ice breakers which are painted red. Black denotes a vessel whose primary mission is aids to navigation and white – like Cuyahoga, denotes a primary mission of law enforcement along with search & rescue.

Commissioned in 1944 Blackthorn had a single propulsion shaft that was driven by an electric motor, which was powered by diesel engines. This propulsion design is referred to as "diesel electric" propulsion. The direction of the shaft rotation could be changed almost instantly merely by changing polarity of the main motor excitation, taking approximately 10 seconds to go from full head to full astern. Engines were controlled through three control stands, one in the pilothouse and one on each bridge wing. The ship's whistle was controlled by a "Y" shaped cable running overhead the width of the bridge and could be sounded from almost any position inside the pilothouse.

She was equipped with a VHF-FM transceiver with a monitor receiver set to Channel 13. This monitor was used to maintain continuous listening guard of Channel 13 for bridge-to bridge communication. Blackthorn also carried 3 handheld VHF-FM radio transceiver units, referred to as COMCO's, for monitoring Channels 13, 16, 21, 22, and 83. When underway, one of these radios, set to Channel 13, was placed on each bridge wing for use by the conning officer.

Homeported out of Galveston, Texas since 1976, she had a compliment of 48 to 50 Coast Guardsmen. She was commanded by Lieutenant Commander George Sepel, a commissioned officer, unlike a warrant officer in the previous two collisions. He had assumed command on the 27th of July 1979.

As with all cutters she was scheduled for a yard availability that would be conducted at a civilian shipyard. This would include extensive work to her machinery and hull and requiring her to be hauling out of the water and placed in drydock.

The facility chosen was the Gulf Tampa Drydock Company in Tampa, Florida. After completing her over 3-month yard availability Blackthorn would head out of Tampa Bay, Florida enroute to her home port of Galveston, Texas with a scheduled brief stop in Mobile, Alabama on the way.

U.S.C.G.C. Blackthorn (WLB-391)

The crew of 50 Coast Guardsmen aboard on the 28th of January 1980 was as follows:

Lieutenant Commander George J. Sepel[49]

Lieutenant Commander George J. Sepel – Aged 34 was married to Georgia, was the commanding officer of Blackthorn, and a 1967 graduate of the U. S. Coast Guard Academy. His previous sea time was the Commanding officer of a 95'patrol boat for 16 months, Executive officer of another 180' buoy tender for 25 months, and 16 months as a deck watch officer on a high endurance cutter.

Author's note – Like CWO4 Robinson in Cuyahoga, LCDR Sepel had been ashore for almost 5 years before assuming command of Blackthorn.

Lieutenant David B. Crawford[51]

Lieutenant David B. Crawford – He was the Executive Officer and a 1973 graduate of the U.S. Coast Guard Academy. He had been on board Blackthorn for roughly 2 years and had prior 180' buoy tender experience.

Ensign Frank John Sarna III – He was 22 years old and a 1979 graduate of the Coast Guard Academy.

Ensign John Ryan[50]

Ensign John Ryan – Aged 29, he was a 1978 graduate of the U.S. Coast Guard Academy and had reported aboard on the 23 July 1979 and became a certified Officer of the Deck in September 1979. He also had previously served in the U.S. Navy.

Chief Warrant Officer 2 Jack Joseph Roberts Jr. – He was 38 years old and enlisted in the Coast Guard on the 28th of January 1964. He was married with a child.

Chief Warrant Officer 2 J.S. Miller – He was the Engineering Officer and reported aboard on the 14th of July 1979.

Chief Machinery Technician David Stidhem

Chief Machinery Technician Luther David Stidhem – He was 39 years old and was married to Darlene Kay Mulhern.

Boatswain Mate Chief Richard O. Robinson

Machinist Mate Chief Ron B. Litterell[54]

Machinist Mate Chief Ron B. Litterell – He was 31 years old and had been in the Coast Guard for 11 years. He was married to Cathi and they had two children.

Quartermaster First Class Jeff L. Huse - He was 27 years old and married.

Storekeeper First Class Ronald G. McCray

Machinery Technician First Class Danny Rinaldo Maxcy – He was 25 years old and enlisted in the Coast Guard on the 3rd of March 1973 and was discharged on the 15th of January 1975. He once again enlisted in the Coast Guard on the 2nd of February 1975.

Machinery Technician First Class Michael LaFond

Machinery Technician First Class Bruce Michael LaFond – He was 32 years old, married, and had 12 years in the Coast Guard.

Substance Specialist First Class Subrino Ibanez Avila – He was 31 years old and enlisted in the Coast Guard on the 5th of May 1967. He was married to his wife Lina.

Electronics Technician First Class Jerome Frederick Ressler – He was 28 years old and had enlisted in the Coast Guard on the 15th of October 1971 and he was married to Renee Elliott Ressler.

Substance Specialist Second Class Clint Campagna

Machinery Technician Second Class Richard Dale Boone[49]

Machinery Technician Second Class Richard Dale Boone – He was 23 years old and enlisted in the Coast Guard on the 24th of Jun 1974 and he was married. He was a native of Modesto, California and a 1975 graduate of Modesto High School

Quartermaster Second Class Gary Wayne Crumly – He was 23 years old and enlisted in the Coast Guard on the 20th of June 1976 and he was married to Glenda. He was a native of Birmingham, Alabama and a 1974 graduate of Erwin High School.

Damage Controlman Second Class Daniel Monreal Estrada – He was married.

Electricians Mate Second Class Thomas Richard Faulkner – He was 22 years old and enlisted in the Coast Guard on the 7th of September 1976 and he was married.

Machinist Mate Second Class P.M. Florence – He was 23 years old.

Substance Specialist Third Class Donald Ray Frank – He was 23 years old and enlisted in the Coast Guard on the 11th of April 1978 and he was married.

Substance Specialist Third Class David B. Marak -He was 22 years old.

Damage Controlman Third Class Lawrence Daniel Frye – He was 21 years old and enlisted in the Coast Guard on the 25th of April 1978.

Quartermaster Third Class Richard Weston Gauld – He was 19 years old and enlisted in the Coast Guard on the 5th of September 1978.

Machinist Mate Third Class Steve G. Overby

Boatswain Mate Third Class Charles E. Bartell – 21 years old from New Orleans.

Hospital Corpsman Third Class Ricky Lee. Chamness – He was 25 years old.

Gunners Mate Third Class Patrick J Lucas – He was 23 years old and from Satellite Beach, Florida.

Electricians Mate Third Class Larry C. Clutter – He was 21 years old from Austin, Texas.

Electronics Mate Third Class Edward Francis Sindelar III[48]

Electronics Mate Third Class Edward Francis Sindelar III
He was 21 years old and enlisted in the Coast Guard on the 7th of August 1978. He was from Greenville, Illinois and a 1978 graduate of Greenville High School.

Fireman Machinery Technician Robert C. Niesel – He was 20 years old.

Seaman Gunners Mate Randolph Brent Barnaby – He was 22 years old and enlisted in the Coast Guard on the 14th of November 1978 and he was married.

Seaman Substance Specialist F.J. Chaplain

Seaman Quartermaster Roger. K. Shine

Fireman D.M. Brooks

Fireman B. J. Dees

Seaman Michael A. Rhodes – From Memphis, Tennessee

Seaman Anthony M. Ware – He was 21 years old and married to his wife Melinda.

Fireman Apprentice Michael Kevin Luke – He was 20 years old and enlisted in the Coast Guard on the 14th of August 1979 and he was married.

Seaman Apprentice George Ronald Rovolis Jr. – He was 17 years old and enlisted in the Coast Guard on the 29th of October 1979.

Seaman Apprentice Glen Edward Harrison – He was 18 years old.

Seaman Apprentice Michael D. Grey

Seaman Apprentice Mark C. Gatz – He was 24 years old from St. Loius, Missouri.

Seaman Apprentice Steven A. Coleman –

Seaman Apprentice E. Solis

Seaman Apprentice John Edward Prosko – He was 19 years old.

Seaman Apprentice Warren Renail Brewer[47]

Seaman Apprentice Warren Renail Brewer – He was 19 years old and enlisted in the Coast Guard on the 29th of January 1979. He was from Memphis, Tennessee and a graduate of Douglas High School.

Seaman Apprentice Charles Douglas Hall – He was 21 years old and enlisted in the Coast Guard on the 27th of August 1979.

Seaman Apprentice William Ray Flores

Seaman Apprentice William Ray Flores – He was 18 years old and from Carlsbad, New Mexico, but later moved to Fort Worth, Texas where he attended Western Hills High School. He joined the Coast Guard in June of 1979.

The following plans of Blackthorn are actually of a slightly more modern 180' buoy tender of the same class. Because Blackthorn is larger than the previous 2 Coast Guard cutters discussed, these drawings are much busier. But I felt they still may help you identify the general layout and spaces you are about to read about.

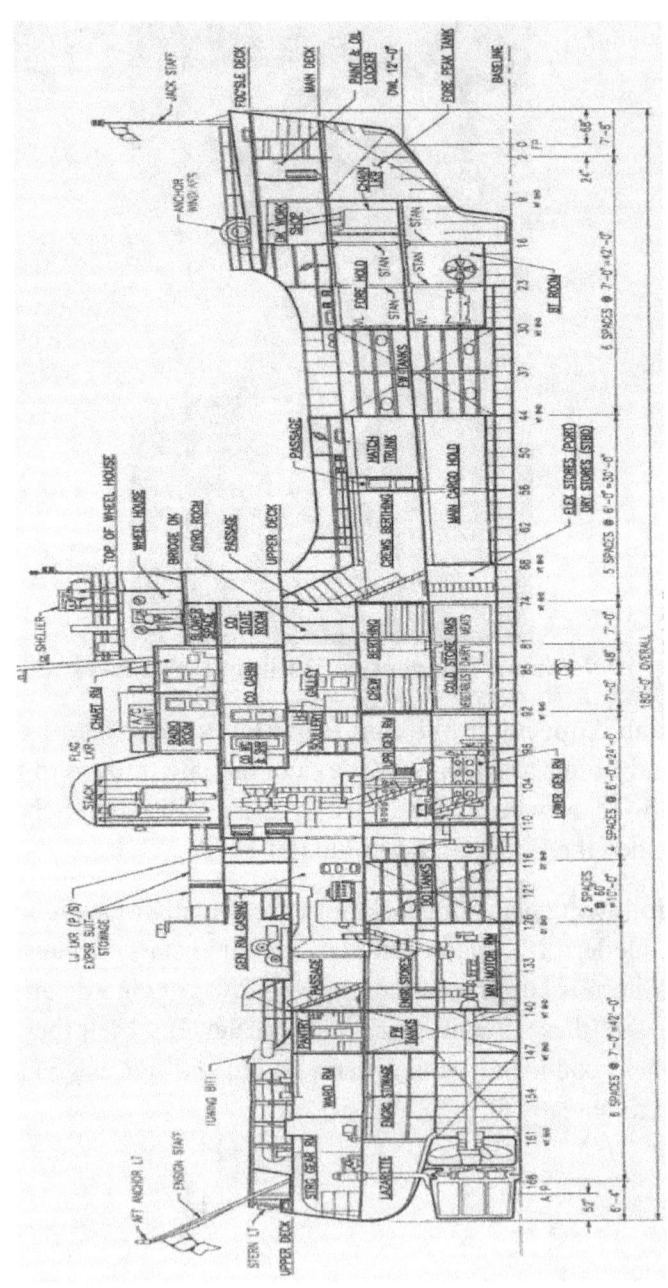

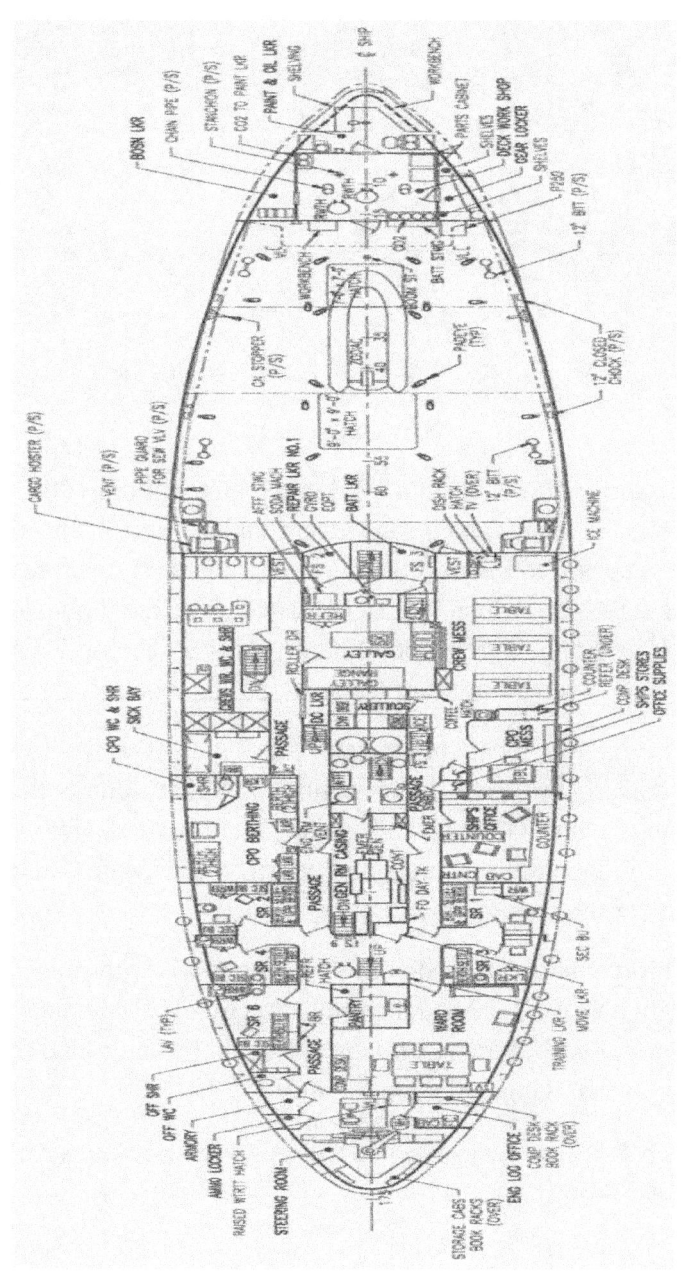

S.S. Capricorn[46]

S.S. Capricorn was built in 1943 and then rebuilt in 1961. She was 605' in length and designed as a tanker. Here propulsion system consisted of two boilers which generated steam to drive a steam turbine, which in turn powered an a/c main generator which in turn powered a 6600-horsepower a/c motor. This single screw could produce a speed of 14.5 knots.

The bridge was located in the aft part of the vessel where her port & starboard – red & green, running lights along with her aft range light -white, and aft whistle were located. Her forward range light – white, and whistle were located on her bow near the anchor windless.

Her bridge was equipped with a gyro compass with repeaters along with a VHF-FM radio with a "piggy back" channel 13 monitor. Her forward & aft whistles could be controlled from either bridge wing and the pilothouse.

On the 28th of January she had a crew of 33. Her known crew was as follows;

Captain George P. McShea Jr. – He was 33 years old and a 1968 graduate of the U.S. Merchant Marine Academy.

Pilot Harry Eugene Knight

Pilot Harry Eugene Knight – He was 38 years old and a 6-year veteran of the Tampa Bay Pilots Association

Chief Mate John Gordon

Second Mate Stephen Sadler – He was the boat officer in the search boat launched by Capricorn.

Third Mate William J. Curran III

Radioman Jack Cohen

Helmsman LeRoy Stoutingberg – He was a veteran of the U.S. Navy.

Merchant Seaman & Cook Robert J. Fitzgibbon – He was 25 years old and on the boat deck helping lower Capricorn's small boat.

Merchant Seaman & Housekeeper Santos Garcia – He was 43 years old and on the boat deck helping lower Capricorn's small boat.

Merchant Seaman O'Boyle – He volunteered to go out in Capricorn's small boat and search for survivors.

Merchant Seaman David Gilmore – He volunteered to go out in Capricorn's small boat and search for survivors.

Merchant Seaman Peter Hulsebosch – He volunteered to go out in Capricorn's small boat and search for survivors.

Merchant Seaman William Thom – He volunteered to go out in Capricorn's small boat and search for survivors.

Merchant Seaman Donald Barney – He volunteered to go out in Capricorn's small boat and search for survivors.

Merchant Seaman Robert Rentz – He volunteered to go out in Capricorn's small boat and search for survivors.

Merchant Seaman Bennie Spencer – He volunteered to go out in Capricorn's small boat and search for survivors.

THE 28TH OF JANUARY

At 1027 Capricorn anchored in the Fairway Anchorage seaward of Egmont Key. The crew of 9 licensed officers, 23 merchant seaman, and one additional person to the crew had just come from St. Croix, Virgin Islands to deliver 151,611 barrels of #6 oil to the Florida Power and Light Company.

Prior to entering the navigable waters of the United States, Chief Mate John Gordon had the steering gear, main propulsion machinery and control communications and alarms tested, but did not ensure that the emergency generator or that storage batteries for the emergency lighting and power systems were tested. He logged that all equipment was tested in accordance with 33 CFR 164.25 – United States Federal Navigation Regulations and reported this to Captain George McShea. It was Captain McShea's policy that all tests required in 33 CFR 164.25 need not be conducted prior to every occasion of entering U.S Navigable Waters, if such tests were otherwise conducted on a periodic basis, such as the emergency generator and storage batteries which were routinely testes once a week, regardless of the vessels location.

At about 1200 Blackthorn set her Special Sea Detail – an all hands evolution for leaving & entering port, and got underway for sea trails to check work completed in the shipyard. She heads out into Tampa Bay and conducts tests well north of the Sunshine Skyway Bridge, which carries highway U.S. 19

between the Pinellas County peninsula on the west and the Manatee County mainland on the east.

While underway an issue with the generators becomes evident and they were not producing enough power. Blackthorn anchored and tried to determine the cause of the problem. Unable to resolve the issue she returned to Gulf Tampa Shipyard to have them look at it.

While the shipyard and Blackthorn engineers worked on the problem, the crew had an early supper, at 1630, in anticipation of getting underway soon.

At 1730 ENS John Ryan piped over Blackthorn's 1MC public address system that the generators were fixed, and Blackthorn would be getting underway in 30 minutes.

At 1804 Blackthorn set her Special Sea Detail once again and departed Gulf-Tampa Drydock in route to Mobile, Alabama with her crew of 6 officers and 44 enlisted crewmembers. The weather was a partly cloudy sky, visibility clear and at least to 7 miles, seas were calm with a light wind out of the north at 5 knots and the air temperature was 61°F and the sea temp was 64°F.

Once in Mobile she was scheduled to load 5 buoys to be set in the Gulf of Mexico and then continue on to her home port of Galveston, Texas.

Prior to getting underway personnel conducted tests of radio and navigational equipment, steering gear, pilot house controls and the ships whistle.

The Blackthorn was fully loaded with 51,270 gallons of water – 95%, and 26,695 gallons of diesel fuel – 95%. Her 16" radar was on and manned for navigational purposes. The 12" radar was operating but was not manned. There were telescopic alidades on each bridge wing gyro repeater to aid in obtaining visual bearings. A VHF-FM COMCO, set to channel 13, was also placed on each bridge wing for the conning officer, who was LCDR George Sepel. Neither of the cutters small boats were rigged for sea and it was LCDR Sepe's policy not to do so.

At 1807 the sun set.

The underway lighting configuration for Blackthorn would be the same as Capricorn. This would be the same lighting configuration as Santa Cruz II and S.S. Helena in the previous stories - for a vessel over 50 meters in length making way. This would once again be, two white range lights – one fore & one aft, and a port red and starboard green light on the pilot house.

Soon after getting underway QM1 Jeffrey Huse had determined the gyro error to be 1° West using a telescopic alidade on one of the terrestrial ranges.

At 1846 Capricorn weighed anchor and proceeded at 70 RPM, 12 knots, into Egmont Channel. The port anchor was housed and set on the brake with the gear disengaged, the devils claw off and the riding pawl down. The starboard anchor was housed and was not ready for letting go. It takes two men to release the anchor windless brake. The port anchor was a 13,500-pound stockless type with eleven shots – 900 feet, of ½ inch stud-link chain.

Author's note – When maneuvering in an area where loss of power could cause a grounding or collision, vessels have one anchor at the ready in case they need to quickly release it to prevent such a mishap.

At 1911 Pilot Harry Knight embarked on Capricorn, after which he assumed the conn and, at the suggestion of Captain McShea, increased speed to 78 RPM – 13.8 knots. Captain McShea didn't pass any other information, nor the status of the anchors. Pilot Knight informed Captain McShea that the Cut "A" Channel Range Front Light – LLNR 1048, was extinguished and that Mullet Key Channel Buoy 14 – LLPG 99, was off station in the direction of center channel.

Author's note – These issues with aids to navigation were normal and the responsibility of a local Coast Guard Cutter similar to Blackthorn to remedy as soon as possible. Also, there are several channels, some referred to as Cut Channels, that merge from inside Tampa Bay to a single Cut Channel, leading in & out of Tampa Bay, which is Cut "A" channel.

Pilot Knight also indicated that there were two piloted outbound vessels that Capricorn would meet on her inbound transit. They were the only piloted vessels of which he was aware. He was familiar with Capricorn and had piloted her on two previous occasions, the most recent being in October 1979.

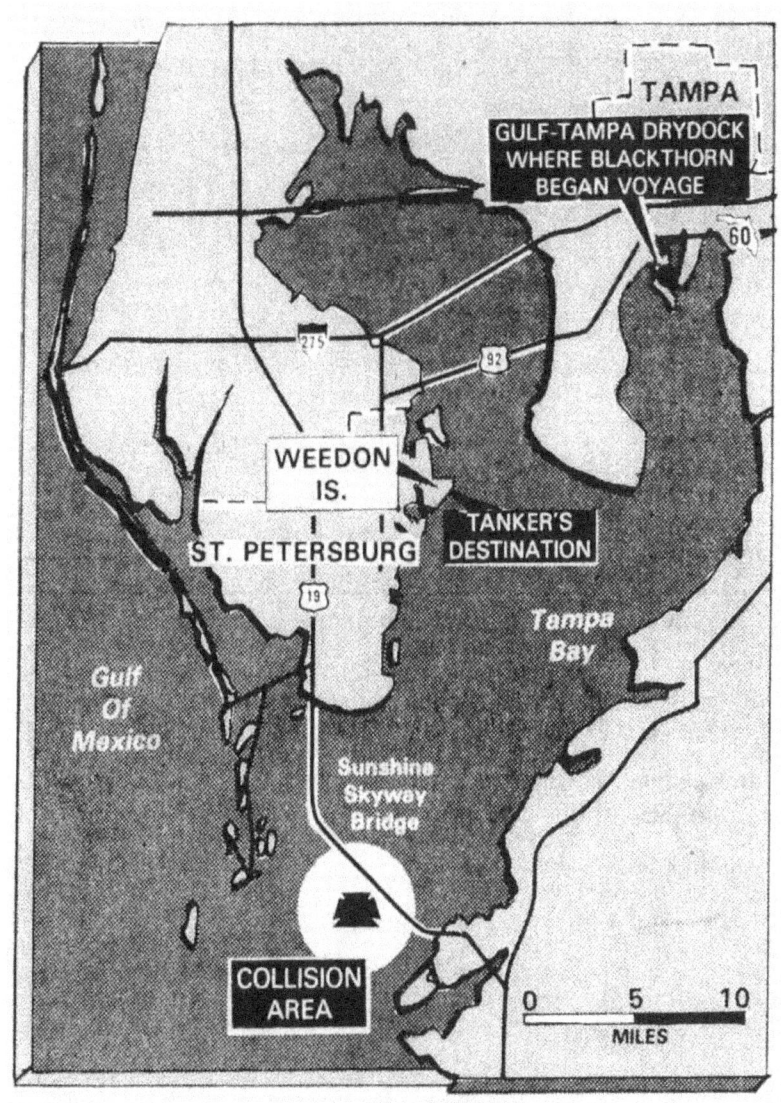

Map of Tampa Bay[52]

As Blackthorn made her way out of Tampa Bay down Sparkman Channel LCDR Sepel observed the Russian cruise ship M.S. Kazakhstan in the Seddon Channel in a southerly direction on a converging course. The brightly lit Kazakhstan was carrying 250 passengers on a 7-day trip to the Caribbean. LCDR Sepel determined that Blackthorn, traveling at approximately 12 knots, was pulling ahead of Kazakhstan and turned into Cut "D" channel ahead of her.

Author's Note – Pilot Gary Maddox, who head the conn aboard Kazakhstan testified that he tried to contact Blackthorn 3 times via Channel 13 but got no answer. He stated that Blackthorn cut ahead of him near Seddon Island.[53]

When clear of Pendola Point and satisfied the Blackthorn was responding well from work completed during the yard availability, LCDR Sepel directed the Executive Officer, LT David Crawford, to assume the conn. During this time, it had been observed that the shaft tachometer was not operating properly. Determining this to be a minor problem, the decision was made to continue and repair it the following day. However, LCDR Sepel decided to go below to the engine room to check out the situation with the Engineering Officer and departed the bridge. Shortly thereafter he returned.

A modified Special Sea Detail was set which retained only the essential personnel for the bridge and engine room watch, along with fully manned anchor and after steering personnel. On the forecastle there were 4 crewmembers ready to let the anchor if needed.

The navigation team on the bridge consisted of ENS Frank Sarna – plotting, ET1 Jerome Ressler – radar, and QM1 Huse – visual bearings. Although the navigation team was plotting periodic fixes and passing them to the conn, the conning officer was using seaman's eye – aka the naked eye, to navigate from one cut to another and employing the terrestrial ranges to help him line up in the channel. Turns were made directly from one channel to another. The "dogleg" wideners were not used. No one was assigned to maintain a shipping plot. The ship's policy was to maintain one only during fog navigation.

During the transit of Upper Tampa Bay Blackthorn met and passed the tugboat Pat B pushing a loaded barge up Cut "E" Channel. Although a port-to-port passage was arranged on Channel 13, Blackthorn maintained a position close to center channel, forcing Pat B to move further to her right – starboard. Blackthorn and Pat B passed at less than 40 feet. The close passing did not affect the maneuverability of either vessel.

Kazakhstan had increased speed and was overhauling Blackthorn. When attempts at radio contact were unsuccessful, speed reductions were necessary so as not to get too close. The pilot aboard Kazakhstan, Pilot Gary Maddox, finally raised Blackthorn on Channel 13. His request that Blackthorn move out of the channel was granted.

Author's note – Pilot Maddox testified that the initial reply from Blackthorn to his request to pass was "Catch me if you can." But minutes later the Blackthorn replied that it would maneuver to the side to let Kazakhstan pass.[53]

The brightly lighted Kazakhstan passed in Cuts "D" and "C" while Blackthorn moved out of the channel and out of her way. Seaman Apprentice Mark Gatz was the lone lookout on the flying bridge wearing a set of sound powered phones. After Kazakhstan passed, he stated, to the Board of Investigation, that his night vision was affected by the bright lights from Kazakhstan.

When Blackthorn reentered in Cut "C" Channel LT Crawford asked LCDR Sepel if he would temporarily relieve him of the conn. LCDR Sepel directed that ENS John Ryan, the Officer of the Deck – OOD, take the conn.

Around 1955, shortly after ENS Ryan assumed the conn, Blackthorn turned into Cut "A" Channel. In doing so ENS Ryan did not utilize the widener or make a "dogleg" turn. ENS Ryan had observed that Cut "A" Channel Range Front Light was extinguished, which he brought to the attention of LCDR Sepel.

As Capricorn entered Egmont Channel Pilot Knight made an open SECURITE call on Channel 13 using his hand-held VHF-FM radio. At that time, he steadied on Egmont Range – LLNR 1032 & 1033 and determined the gyro error to be zero by making an observation on the range without benefit of a bearing circle. Prior to this, Capricorns gyro error had usually been 1° West according to Captain McShea.

At 2000 Capricorn passed Egmont Key Light – LLNR 1038, abeam to starboard, proceeding towards Mullet Key Channel. Pilot Knight observed the outbound Kazakhstan east of the Sunshine Skyway Bridge in the vicinity of Cut "B" Channel, approximately 7 to 8 miles away. At his time, he received an open SECURITE call from the tug Ocean Star, which was pushing a loaded barge in Mullet Key Channel. Pilot Knight set a course of 081°T upon entering Mullet Key Channel and steadied up on the range.

In addition to Pilot Knight and Captain McShea the bridge was manned by the 2000-2400 watch which consisted of a Mate of the Watch, a Helmsman, and a lookout stationed on the bow. The lookout did not have binoculars and was instructed to only report vessels that seemed unusual. The 10cm radar was operational and was looked at periodically by Captain McShea and the Mate of the Watch. The 3cm radar was on standby.

Capricorn met and passed the tug Ocean Star just east of Mullet Key Channel Buoy 15 – LLPG 99, and Lighted Buoy 16 – LLNR 1045. At the time of this passing, Capricorn was close to or on the range. Soon after, Pilot Knight observed Kazakhstan outbound, doing excess of 14 knots turning into Mullet Key Channel from Cut "A." The bow lookout called the bridge to report this vessel because he felt that it was in the middle of the channel and that it was going to be a close passage. Radio communications were established with Pilot Maddox on Kazakhstan using the pilot's handheld radio on VHF-FM Channel 13. Both Pilot Maddox and Pilot Knight agreed to a port-to-port passing. According to personnel on the tugboat Pat B and Blackthorn, who overheard the conversation, Pilot Maddox also advised Pilot Knight that a Coast Guard vessel was following Kazakhstan.

Author's note – Captain Russell Wanamaker and Kenneth Baker, an able-bodied seaman aboard Pat B, stated to the Board of Investigation that they heard the pilot aboard Kazakhstan tell the pilot aboard Capricorn that there was a Coast Guard vessel behind them. QM1 Huse also stated that he heard the Kazakhstan brief someone that a Coast Guard vessel was behind them. However, Pilot Knight and Pilot Maddox testified that this conversation occurred after the collision.

After the exchange the two pilots then shifted to VHF-FM channel 77 for continued discussions. As the two vessels passed close aboard, Pilot Maddox apologized to Pilot Knight for crowding him. After passing Kazakhstan, approximately 50 feet to port, Pilot Knight, still talking to Pilot Maddox on channel 77, left the port bridge wing and went to the starboard bridge wing to observe Mullet Key Range. Captain McShea followed Pilot Knight inside the pilothouse and stopped at the radar. At this time Pilot Knight first observed the white masthead and range lights and green side light of Blackthorn as she proceeded outbound west of Sunshine Skyway Bridge in Cut "A." Although Captain McShea had periodically checked the radar, which was on the 3-mile scale, he had not previously detected this outbound contact. He now sighted the Blackthorn approximately 2/3rds of the way down Cut "A" Channel from the bridge. The bow lookout had sighted Blackthorn ahead but did not report it because he didn't consider her position unusual and was instructed not to report well-lit vessels in order to keep voice traffic to the bridge to a minimum.

As Blackthorn approached Sunshine Skyway Bridge, SA Gatz was a lookout on the flying bridge and the anchor detail on the forecastle was considered by LCDR Sepel as a second lookout.

Author's note – Members of the anchor detail testified that they did not consider themselves as lookouts.

SA Gatz had been on watch for over 2 hours. He was wearing sound powered phones and was on the 1JV circuit along with SA Steven Colman on the forecastle and HM3 Ricky Chamness on the bridge. ENS Ryan directed an additional circuit to be manned when he learned that phone talkers were experiencing background noise on the 1JV circuit. SK1 Ronald McCray was called to the bridge and directed to man the 2JV circuit along with SNQM Roger Shine in after steering and an unknown person in the engineroom. Extra personnel were in the engine room to monitor recent work that was performed during the yard period. Although there was also background noise on the 2JV circuit, communications were possible on both circuits.

As Blackthorn approached Sunshine Skyway Bridge, SA Gatz became focused on the 87' shrimp boat Bayou, who appeared to be overtaking Blackthorn. The navigation team, under the direction of ENS Sarna, was locating new terrestrial points to use for fixes after passing under the bridge. LT Crawford had returned from below and was on the port bridge wing with LCDR Sepel and ENS Ryan. Since Cut "A" Range Front Light was extinguished, ENS Ryan was periodically taking visual bearings on Cut "A" Range Rear Light to help determine his position with respect to the center of Cut "A" Channel. All of the bearings were virtually constant with no appreciable change as they moved down the channel. The course written on the navigation chart was 242°T. The range line for the channel is 243°T.

ENS Ryan conned Blackthorn under the Sunshine Skyway Bridge just to the right of the green navigation lights positioned on the underside of the bridge marking the centerline of Cut "A" Channel. At that time ENS Sarna marked the vessel's position on the navigation chart as minute 18, which indicated a speed of 11.7 knots. LCDR Sepel went inside the pilothouse, looked at the 16" radar, and observed a large radar contact to the east of Mullet Key Channel Buoy 15. He looked ahead and saw the bright lights of Kazakhstan on the same relative bearing as the radar contact. LCDR Sepel then moved to the chart table to examine the chart and refresh his memory on the trackline past Egmont Key. Soon after clearing the Skyway Bridge, LT Crawford asked ENS Ryan if he had talked to a vessel whose navigation lights were just beginning to separate from the lights of Kazakhstan in Mullet Key Channel – tis would be Capricorn. ENS Ryan replied that he had not. He requested that LT Crawford contact the inbound vessel, Capricorn, on Channel 13 using the COMCO VHF-FM transceiver so that he could concentrate on evaluating the oncoming contact. HM3 Chamness while inside the pilothouse also sighted lights emerging from behind Kazakhstan. After watching them for a period of time and noting that they appeared to have a very rapid left – port, bearing drift, HM3 Chamness asked the flying bridge lookout, SA Gatz, if he saw them. When SA Gatz verified that he did see the lights. HM3 Chamness did not report the sighting to the conn because it appeared by then that ENS Ryan was already aware of them. The anchor detail on the bow had their attention concentrated on the Kazakhstan ahead of them. The passenger liner was brilliantly lit with red, green, blue, and white deck lights and there was a festive atmosphere about her.

LT Crawford called the inbound vessel in Mullet Key Channel and a short time later heard a garbled response, followed clearly by the words, "coming out of anchorage, won't be in your way." LT Crawford testified that LCDR Sepel ordered him to "roger" for the transmission, but LCDR Sepel in testimony denied giving such an order. There were no other witnesses who recalled such an exchange. At approximately the same time, Kazakhstan was conversing with Ocean Star on VHF-FM Channel 13 in the vicinity of Mullet Key Channel Lighted Bell Buoy 13 LLNR 1044. Ocean Star, using 25-watt power, was informing Pilot Maddox on Kazakhstan that she was crossing the channel to enter the anchorage east of Egmont Key and that she would be out of his way and leave him plenty of sea room.

On Capricorn Pilot Knight shifted his radio back to Channel 13 and attempted to contact the outbound Blackthorn with no reply.

Author's note – Pilot Maddox testified that he heard Pilot Knight on Capricorn call Blackthorn as he was attempting to call Ocean Star.

Pilot Knight walked back inside the pilothouse and stood just to the left of the centerline of the ship. He observed the relative bearing of Blackthorn and the aspect of her running lights remained constant for about 30 seconds. The bearings were taken visually without benefit of a bearing circle. He attempted a second call and again received no reply. Pilot Knight, constrained by Capricorns draft to remain in the channel and approaching his normal turning point, did not change course or speed. He decided against sounding a one short blast whistle signal proposing a port-to-port passage, because he did not want to limit the options of the other vessel. By this time Blackthorn appeared to have passed Lighted Buoy 1A. Now, Pilot Knight felt Blackthorn was not going to change course into Mullet Key Channel but, instead, was going to cross ahead and go out the channel east of Lighted Bouy 2A.

Author's note – Expecting Blackthorn to exit the channel at this location would have been normal because it is where the Intercoastal Waterway merged with Cut "A" Channel.

At this time Captain McShea, concerned with the lack of radio contact and trying to evaluate the situation, left the radar and walked up alongside Pilot Knight and said, "What's this guy trying to prove?" Pilot Knight did not respond to the comment or indicate the Intercostal Waterway and Southwest Channel traffic departed Cut "A" Channel near Lighted Buoy 2A. By this time the bow of Capricorn was fast coming up abeam of Lighted Buoy 2A. Capricorn had to commence its turn – to port, soon or risk grounding on the southeastern bank of Cut "A" Channel.

LCDR Sepel was still at the chart table when he overheard LT Crawford's radio call and the garbled response, "going into or coming out of anchorage – he couldn't remember which, I won't be in your way," over the Triton monitor. This alerted him to the presence of another vessel in addition to Kazakhstan. In the meantime, ENS Ryan had returned to the port bridge wing gyro repeater after passing through the pilothouse and stopping to take a short look at the radar. He took a second bearing on the incoming vessel, determined a slight left bearing drift, then turned and took a quick bearing astern on Cut "A" Range Rear Light and found it to be the same as the previous ones. Having heard nothing from LT Crawford, and assuming that a port-to-port passage had been successfully arranged, ENS Ryan decided to commence his right – Starboard, turn into Mullet Key Channel early in order to provide a little more sea room for the oncoming vessel, Capricorn. He ordered helmsman, QM3 Richard Gauld, to come right – Starboard, to 263°, which was the gyro compass course for Mullet Key Channel. He did not give a rudder command, leaving the amount of rudder to the helmsman.

ENS Ryan did not take any turn bearings or a bearing on Lighted Bell Buoy 1A to see how this new course change would leave the buoy. He was using seaman's eye to negotiate the turn. Also, he did not sound a one short blast whistle signal due to the ship's general policy of not sounding whistle signals if passing agreements had been made by radio. LCDR Sepel felt this policy would avoid possible confusion. LCDR Sepel then stepped onto the port bridge wing, sighted the oncoming Capricorn for the first time, and said "Where the fuck did he come from?" he motioned to ENS Ryan to continue right – Starboard.

Pilot Knight now felt that there was no chance for a port-to-port passing. He put the rudder left – port, 10° and sounded two short whistle blasts. Captain McShea had reached the same conclusion and silently concurred with Pilot Knights actions.

Author's note - Captain McShea testified that he would have immediately taken over the conn if he had not thought it the proper thing to do. He said "right then and there I figured this guy is crossing our bow. I had him on our side of the channel and I had his starboard green light in view."

No answering whistle signal was received from Blackthorn.

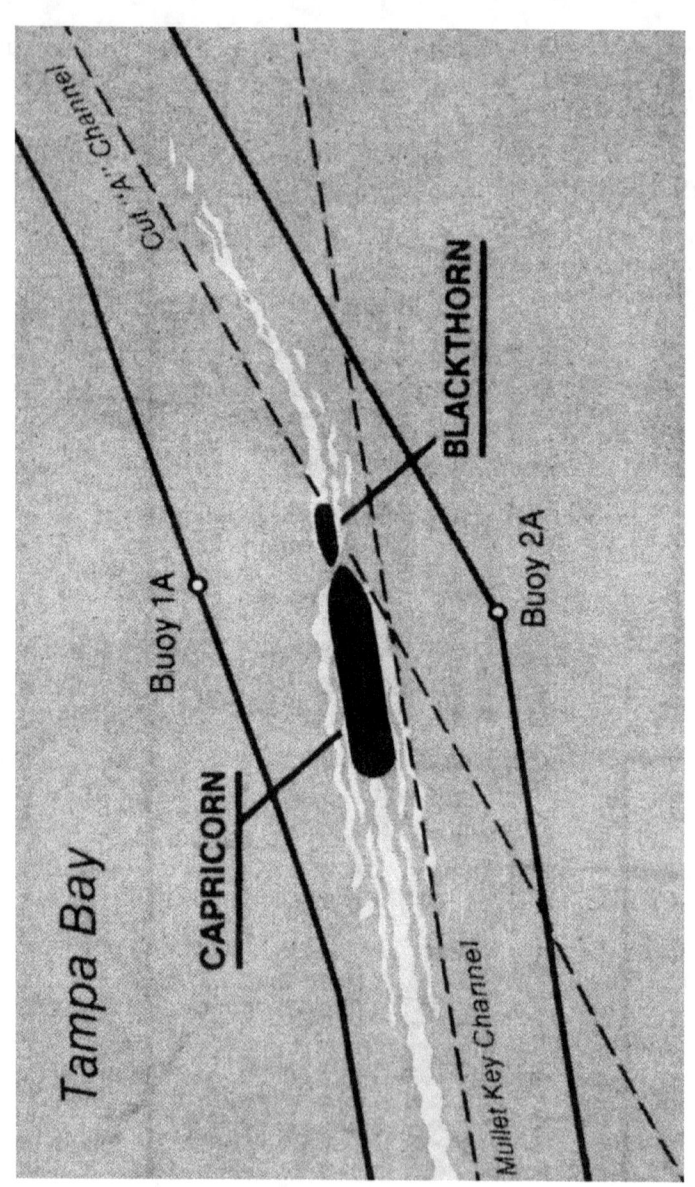

Map of positions just before the collision as per the Board of Investigation[61]

Author's note – You can see in the drawing that Buoy's 1A & 2A are at the bend that ends Mullet Key Channel & where Cut "A" Channel begins. The channel also narrows here from 600' down to 400' for inbound vessels and widens for outbound vessels. The center track line that exits the channel from Cut "A" Channel leads to the Intercoastal Waterway. This is a depiction of where the Board of Investigation said the collision occurred.

LCDR Sepel heard the whistle signal from Capricorn and yelled "Right full rudder!", thereby automatically taking the conn from ENS Ryan. This order was heard by the bow lookout on Capricorn.

Pilot Knight ordered the rudder 20° left – port, then immediately hard left, and sounded the danger signal. Captain McShea then observed the range lights of Blackthorn line up very quickly, indicating a right turn directly towards Capricorn.

Autor's note - Captain McShea testified that "all at once he – Blackthorn -came hard right and hit us on the port bow."

LCDR Sepel on Blackthorn ordered that "Standby for collision" be piped and put the engines back full, using the port bridge wing pilothouse control. Blackthorn's course had changed about 20° to the right – starboard, when the impact occurred. All anchor detail personnel scrambled off the bow onto the buoy deck just before impact except for SN Michael Rhodes, who had previously relieved SA Colman as phone talker. He couldn't unhook the headset in his haste and thus was trapped on the bow at impact.

At 2021, according to Capricorns bell book, Capricorn and Blackthorn collided port bow to port bow at a relative angle of 180°. It is estimated that the collision occurred between 5 to 15 seconds after Capricorn sounded the danger signal.

Immediately after the impact Capricorn stopped her engines. The collision forced Blackthorn to an approximate 15° starboard list, after which she rolled to port and settled at an approximate 5° port list. As both vessels continued past each other, Capricorn's port anchor raked Blackthorn's side and ripped into the crew's head and shower area, where it became imbedded as the anchor chain commenced running out. Capricorn placed her engines full astern with her rudder remaining hard left – port. Blackthorn's engines had been backing full since before impact and her rudder remained right full.

Capricorn decelerated while in a left – port, turn, towing Blackthorn stern first in Cut "A" at a high rate of speed. Pilot Knight kept the rudder hard left to ground Capricorn and avoid colliding with the Sunshine Skyway Bridge. The anchor chain continued to run out, overriding the brake. The bow lookout on Capricorn had watched the two vessels collide until dust and debris from impact and the running anchor chain clouded his vision. He also heard the port anchor chain running out immediately after impact.

About 20 off duty crewmembers assembled on Blackthorn's messdeck, the mustering location of the damage control party, and awaited further instruction. The General Quarters alarm had not been sounded and no further instructions had been piped over the 1MC public address system. CWO2 John Miller, the Engineering Officer, determined from MKC Luther

Stidhen, the Engineering Officer of the Watch, that the engine room was not taking on water. CWO2 Miller checked the motor room and found it dry also. Three individuals who had been in after steering exited that compartment, dogging the door behind them, but leaving the port side porthole open. BM3 Charles Bartell, after emerging from forward berthing, dogged the watertight doors to forward and after crews berthing.

There was confusion and panic on the messdeck. New crewmembers who had reported on board during the yard period froze. During the nearly 3-1/2-month yard period, six new crewmembers had reported aboard straight from boot camp or via class "A" school.

Author's note – Boot camp is initial entry training into the Coast Guard and class "A" school is a term used to describe the Coast Guards job rating school, which generally occurs within the first year of enlistment.

This was their first day underway on Blackthorn. Some of them went to the Watch, Quarter and Station Bill posted in the starboard passageway aft of the messdeck, adjacent to the ship's office, to check their collision at sea assignments. BM3 Bartell took charge on the messdeck and ordered material condition Zebra set, including the closing of the open portholes. He also ordered damage control equipment broken out from the repair party locker and personnel to report to their General Quarters stations. SS2 Clint Campagna reported to the bridge by telephone from the messdeck and was told to standby for instructions.

MK1 Bruce LaFond, who had been in the portside crew's shower area at the time of the collision, was found by MKC Rondal Litterel and CWO2 Miller in the port maindeck passageway, naked, injured, and in apparent shock, mumbling something about the anchor in the shower. MKC Litterel went forward to the crew's head and shower area to check out the damage, and found a large anchor imbedded in the joiner bulkhead separating the crew's shower from the vestibule at the top of the port ladder from the after crews berthing. SN Rhodes, having been injured on the forecastle at impact, was being helped to the messdeck by SS3 David Marak and SN Anthony Ware. FNMK Robert Niesel, who had been in after berthing when the collision occurred, was on the messdeck securing portholes when he became concerned that someone might still be in after berthing. He undogged the watertight door, 1-77-1, and went below and thoroughly checked the space, finding no one.

On the bridge, QM1 Huse was directed to broadcast a Mayday on VHF-FM Channel 16. Scott Hutchins, a Coast Guard duty officer at Coast Guard Group St. Petersburg, Florida received the "Mayday" call at 2020 stating that Blackthorn had been in a collision and taking on water. Group St. Petersburg requested a state of damage and Blackthorn replied, "Standby," at 2023. That was the last Group St. Petersburg heard from Blackthorn. At about the same time Capricorn transmitted a SECURITE call on channel 16 stating that it had been in a collision with another vessel, which appeared to be sinking. Group St. Petersburg requested more information and an unidentified vessel, assumed to be the shrimp boat Bayou, broke in at 2025 and reported a vessel had sunk and asked for assistance on scene.

Author's note – The official transcript was as follows:[58]

Blackthorn, 2022 – Mayday, Mayday, Coast Guard Group St. Petersburg, Group St. Petersburg. Coast Guard Cutter Blackthorn, Cutter Blackthorn, Channel 16 over.

Group St. Petersburg – Cutter Blackthorn, Coast Guard Group St. Pete, over.

Blackthorn – Cutter Blackthorn, Group St. Pete (pause). This is Cutter Blackthorn. Be advised we had a collision. A collision to the seaward side of the Skyway Bridge. Proximate position one alpha, Mullet Key Channel, over.

Group St. Petersburg – Cutter Blackthorn, Group St. Pete, roger, roger. Request to know if you are taking on any water and how bad the damage is, over.

Blackthorn, 2023 – This is Blackthorn, standby, standby this channel.

Capricorn, 2024 – Security, security, security. Tanker Capricorn, KIXX, west of the Skyway Bridge, just had a collision with another vessel which appears to be sinking. Security, security, security, security.

Group St. Petersburg – Coast Guard Group St. Pete back to tanker involved in collision. Please come back with your name over.

Capricorn – This is tanker Capricorn. How do you read?

Unknown Vessel – Don't worry about how you read. This tug is sunk here. You better get some help.

Coast Guard St. Petersburg – Captain what's the situation out there? Over.

Capricorn – It looks like a sunk vessel. Dispatch all vessels. All boats to Skyway Bridge.

Group St. Petersburg – Group St. Pete roger. The Capricorn, was that vessel involved a Coast Guard cutter? Over.

Capricorn – I believe so Captain. The best I can tell it was. Roger.

Group St. Petersburg – Roger, and you are the vessel that was involved in the collision?

Capricorn – That is correct Captain.

Group St. Petersburg – Roger, have you sustained any damage to your vessel? Over.

Capricorn, 2025 – We might be grounded, but we are not sure now. Standby.

Group St. Petersburg – Roger

Group St. Petersburg – Coast Guard Cutter Blackthorn, Coast Guard Cutter Blackthorn, this is Coast Guard Group St. Petersburg. Over.

Unknown vessel – The Blackthorn was the one that put out the mayday.

Group St. Petersburg – Mayday relay, Mayday relay, Mayday relay. All shipping Tampa Bay, all shipping Tampa Bay. This is

Coast Guard Group St. Petersburg. Any vessels Tampa Bay requested to proceed and assist the sinking Coast Guard cutter Blackthorn. Position seaward, west of the Skyway Bridge. All vessels in the vicinity are requested to proceed and assist Coast Guard cutter Blackthorn suspect sinking west of Skyway Bridge. This is Coast Guard St. Petersburg.

QM2 Gary Crumly went to the chart room to break out the inflatable lifejackets for the bridge personnel. LCDR Sepel asked ENS Sarna where the nearest shoal water was and received a reply of Mullett Key Shoal to the north. At that time LCDR Sepel brought the engines to stop and then may have placed them ahead to proceed towards shoal water. There is conflicting evidence as to whether or not LCDR Sepel actually put the pilothouse controls in the ahead position. The first divers on scene found the pilothouse controls at the stop position. The malfunctioning tachometer inside the pilothouse showed that the engines were astern at approximately 100 RPMs when the gauge stopped functioning after capsizing. Although LCDR Sepel could not remember putting the controls ahead during his testimony, his personal notes written hours after the collision stated he had done so. Many survivors reported seeing the propeller still turning after the vessel capsized, but the only one who could recall the direction of rotation described it as being in a backing direction.

Blackthorn suddenly rolled to port and capsized within 15 to 20 seconds. LCDR Sepel, while still handling the pilothouse controls on the port bridge wing, shouted, "Abandon ship!" no one had time to pipe the order over the 1 MC public address system before the vessel rolled on her port beam. As the vessel rolled over, the ship's service generator tripped offline, and the

Blackthorn's lights went out throughout the ship. No personnel could recall any of the emergency lanterns coming on. Personnel on the messdeck exited fore and aft. BM3 Bartell attempted to reach the starboard boat to cut the boat falls, free it, and get it over the side. By that time the vessel was too far over, so he continued aft towards the fantail in an unsuccessful attempt to cut the life rafts free before they were submerged. EM3 Larry Clutter ran to the 01-deck starboard lifejacket locker and started throwing lifejackets to men already in the water. Several men who were fortunate enough to reach the starboard side of Blackthorn as she rolled on her port side climbed the hull onto the keel as the vessel continued to roll to an inverted position.

SA Gatz on the flying bridge stated "There was a hesitation, just after the crash, and I got up and looked around. I ran over to the port side and saw it was entirely caved in. I tried to contact the bridge but there was confusion, too much going on, I was terrified." After getting no answer from the bridge he tried to contact it again. After getting no answer he signed off with "Flying Bridge offline" and Blackthorn took her fatal roll. He went on to say that "I tried to get the phone jack out of the box, but the ship took a tilt to port. I was holding onto the rail to keep from sliding off the flying bridge. Then the Blackthorn rolled until its port side was flat in the water and I found myself vertical, hanging on."

As Blackthorn began to sink, he began to be dragged down with her because of his sound powered head set still being connected to his head. He pulled out his pocketknife and cut at the chin strap of the sound powered phones. As he slashed at the chin strap he cut his cheek, but he was free and pushed himself off to the surface. After surfacing he clung to a board with 4 other shipmates. He later grabbed a life vest for added insurance. Besides the cut on his cheek, he had swallowed some contaminated seawater with fuel and oil.

FNMK Niesel had just finished his inspection of after berthing when Blackthorn rolled. He fought his way up the starboard ladder as water rushed up the athwartships passageway and flooded after berthing. Upon reaching the top of the ladder the flooding water swept him onto the inverted messdeck, where he swam through the starboard watertight door to the buoy deck and escaped.

As Blackthorn rolled to an inverted position, SA Warren Brewer climbed up into the engine room through the escape scuttle located aft on the messdeck. He shined a flashlight down through the scuttle and shouted "I've found a way out! I've found a way out!" Shipmates, still trapped in the pitch dark and rapidly flooding messdeck rushed for the scuttle and apparent safety. SA Gray quickly realized that the engine room was not the way out, convinced SA Charles Hall and SN Rhodes to follow him. They linked themselves together and followed SA Gray towards the starboard door which led out to the buoy deck. When they reached the door, SN Rhodes went forward and standing on the port bulkhead of the starboard vestibule, opened and held the 130-pound starboard watertight door, 1-70-1, over his head while Blackthorn was on her port side so that his 2 shipmates could escape. Then, he escaped as water rushed in. As they exited the messdeck the three men became separated, and SA Gray surfaced alone beneath the inverted buoy deck, were he encountered LCDR Sepel and reported the situation. Realizing that the rapidly flooding vessel was in imminent danger of sinking, LCDR Sepel ordered SA Gray not to attempt to return to the messdeck. He directed SA Gray to dive under the bulwark and swim away from the vessel to avoid being carried under by suction of the sinking Blackthorn.

Blackthorn's inport quarterdeck shack had broken free from the vessel and crewmembers mustered around it as it floated near Blackthorn. There, CWO2 Miller cajoled and organized them. He ordered other men off the capsized Blackthorn and to don lifejackets and he mustered them around the quarterdeck shack for mutual support while awaiting rescue. The unconscious body of CWO2 Jack J. Roberts Jr. was brought to the quarterdeck shack by FNMK Niesel and EM3 Clutter. They along with CWO2 Miller and BM3 Bartell kept CWO2 Roberts' body afloat.

SA William Flores was last seen on board Blackthorn throwing life jackets to his shipmates who were already in the water.[60]

All of the survivors reported difficulty in being able to make full use of their life jackets. Many of them could not untie the tightly bundled jackets in the cold water and clutched them to their chest as a float. Others, who were able to don a jacket, either had difficulty in finding the securing straps and left the jacket hanging open or, if unable to sort out the various straps, grabbed them into a clump and tried to hold the jacket together. In no case did any of the survivors utilize the leg straps and in only one case were the collar straps tied together. One survivor put his head through one of the armholes and was able to utilize it in that fashion.

Capricorn's anchor fell from Blackthorn's side to the channel bottom, about 527 yards from the junction of Mullet Key Channel and Cut "A" Channel range lines. A section of Capricorns bolster and an 11' by 3' section of steel plate weighing about 1100 pounds was found in the same location. Capricorn's bow ground on the north side of Cut "A" Channel, 800 yards from the junction of Mullet Key Channel and Cut "A" range lines. She then pivoted left – port, finally coming to rest on a heading of 281°T. She reported on Channel 16 that she was aground.

After Blackthorn capsized there was free access for water entry above the main deck through the port side shell damage from frame 68 to frame 98. Flooding occurred through the port side hull damage below the main deck from frame 58 to about frame 66. This small penetration opened the hull in a way of the main hold and forward berthing. Flooding to after berthing would have also occurred through the damaged watertight door structure, 1-89-2, and vestibule at frame 89, port side.

Watertight access to the engine room, motor room and after steering were secured. The after steering porthole, port side, was open at the time of the collision and was not secured after the collision. Flooding of the engine room and motor room would have occurred through vents on the 01 deck at frame 115, after Blackthorn rolled 90° and the water level was above the centerline.

The shrimp boat, Bayou, which had been following Blackthorn in Cut "A" Channel, arrived on scene within minutes and commenced rescuing survivors. She had a crew of three, Captain Bill Parker, Vince Dyer & Nick Whitelaw.

The Bayou crew said as they came around Capricorn, they could see the Blackthorn on its port side sinking and her crewmen thrashing around in the water.

Oil was everywhere and there was a strong current pushing the floating crewmen towards the Sunshine Skyway bridge. They said the water was littered with empty lifejackets.

As they tried to hoist the survivors aboard, they said they were slick like noodles from the oil & fuel in the water and hard to get aboard. Once on board they complained that their skin and eyes were burning from the oil & fuel and were given showers.

One of the crewmembers was so slick and shirtless, they couldn't get him aboard - they were referring to CWO2 Roberts. Nick Whitelaw stated "One guy, we just couldn't get on board because he was so oily. He didn't have a shirt on, and he was so slick there was no way to grip him. We held him for a long time, Vince and me. He didn't seem to have a pulse. Finally, we got him tied to some Coast Guard boat."

Author's note – The Coast Guard boat was CG-41452.

The crew of the Bayou stated that the Blackthorn survivors stated that they thought the collision alarm was a drill. In all,

23 of the 27 surviving Blackthorn crewmen were rescued by the Bayou and her crew of three.[55]

Group St. Petersburg had ordered several Coast Guard small boats to head to the scene and had recalled the Coast Guard Cutters Vice (WLIC-75305) & White Sumac (WLM-540) to also get underway. Along with active-duty Coast Guard assets Coast Guard Auxiliary, Eckerd College Search & Rescue Unit,

Pinellas County Sheriff's Office, St. Petersburg Police Department, Tampa Fire Department, Palmetto Police Department, commercial shrimp vessels and civilian volunteers all headed to the scene of the collision. There was a total of 10 vessels that arrived on scene.

Airborne helicopter support was launched by the Coast Guard, Pinellas County Sheriff's Office, Tampa Police Department and St. Petersburg Police Department.

CG-41452 arrived on scene and rescued the remaining 4 survivors. Trying to get the body of CWO2 Roberts was also proving difficult to the crew of CG-41452 and several Blackthorn crewmembers aboard the Bayou transferred over to CG-41452 to assist in getting his body out of the water. BM3 Bartell eventually reentered the water and assisted in placing CWO2 Roberts body in a stokes litter and he was lifted aboard CG-41452. Once out of the water BM3 Bartell commenced resuscitation efforts on lifeless CWO2 Roberts under the direction of HM3 Chamness in a vain attempt to save his life. He was later pronounced dead, and his cause of death listed as drowning. He was buried at Machtelah Cemetery in Pascagoula, Mississippi.

Capricorn now grounded, launched her lifeboat with a crew of 7 volunteers within 24 minutes and continued to search for survivors for 2 hours. On the boat deck Merchant Seaman Robert J. Fitzgibbon Jr., who was a cook aboard Capricorn, said he could hear cries for help as he lowered the lifeboat with Merchant Seaman Santos Garcia. Garcia stated that he heard Blackthorn crewmen shouting for help and saw Blackthorn drift past the Capricorn wallowing from side to side then capsized and sinking stern first.

Civilian divers from Eckerd College Search & Rescue Unit arrived on scene at around 2230 and conducted scuba diving on Blackthorn in an attempt to detect the presence of survivors. They found Blackthorn about 50' down, resting on her port side. They did not enter the sunken vessel and no evidence of life was detected. They also noted that they found working in the Bay very difficult because of the current.

All 27 survivors were transported to shore rescue and assistance personnel at Fort De Soto located at Mullet Key Bayou. It would also be the sight of a temporary tent shrouded morgue.

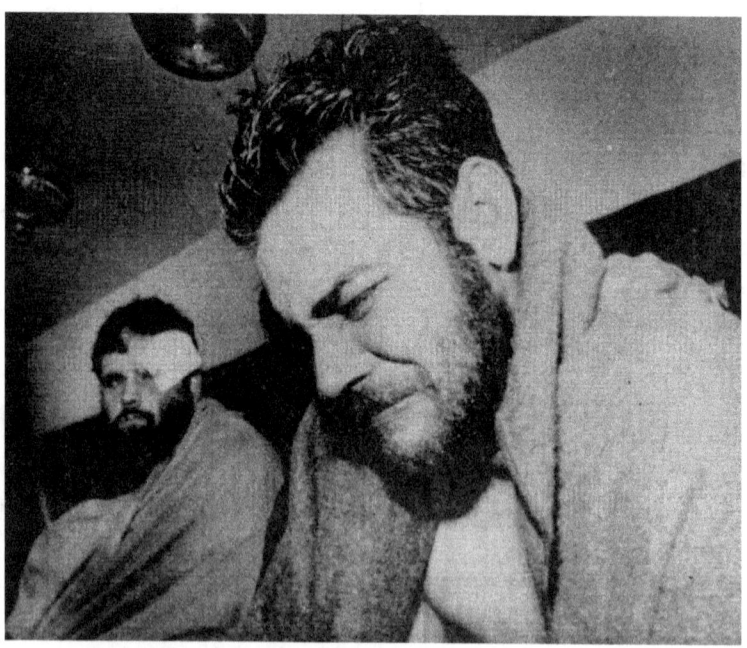

Blackthorn Survivors after being rescued[59]

AFTERMATH

At 0500 on the 29th of January, 3 hard hat divers from Coastal Diving and Marine Construction of Tampa arrived on scene. They dove two times for orientation and to help position the USCGC Vise over Blackthorn. They said the lifeboats were upside down and still in their davits and that there was a lot of debris. It was also dark & murky.

On the third dive the starboard side of the hull was tapped on from the outside with their knives, but they couldn't hear a reply. They were hoping that survivors may have found an air pocket in a compartment and were awaiting rescue. Debris could be heard randomly striking the hull from inside of Blackthorn. On the 4th through 6th dives 3 bodies were recovered.

At 0802 the first body was brought to the surface and was examined by Coast Guard Commander Robert Pettyjohn, who was a physician. Coincidentally, he was also involved with examining the victims of the Coast Guard Cutter Cuyahoga collision. When a body was brought up to the surface, Commander Pettyjohn inspected the body and documented any bruises, distinguishing marks or jewelry.

MK1 Bruce M. Lafond was recovered in the after area of the messdeck. His death certificate lists his cause of death as drowning. He was buried at Galveston County Memorial Cemetery in Hitchcock, Texas.

MK1 Danny R. Maxcy was recovered at the base of the upper ladder to the engine room. His cause of death was listed as drowning. He was buried at Spring Hill Memorial Gardens in Mobile, Alabama

SS1 Subrino I. Avilia was recovered at the hatch leading from the aft athwartships passageway to the fantail. His foot was caught on the wire ladder rail and a strap of the lifejacket he had partially donned was snagged on something in the hatch. His cause of death was listed as drowning. He was buried in Galveston, Texas.

As the divers brought the first three bodies to the surface, they commented on the unusual fact that they appeared to all be first class petty officers.

Author's note – Newspaper coverage of the divers bringing bodies to the surface was graphic and showed the lifeless crewmen.

The divers also didn't find any air pockets or survivors inside areas they could access inside Blackthorn.

At the same time Pinellas Park Fire Department divers retrieved two bodies from the buoy deck area.

SA Charles D. Hall was wedged between the starboard boom vang winch and the bulkhead at frame 70. His cause of death was listed as drowning. He was buried at Culpepper National Cemetery in Culpepper, Virginia.

ET1 Jerome F. Ressler was found on the forward port side of the buoy deck. His cause of death was listed as drowning. He was buried at Arlington National Cemetery in Fort Myer, Virginia.

After going aground Capricorn dropped her starboard anchor to secure her position. All spaces were checked for damage and no holed tanks were found. Fearing the port anchor might still be snagged in Blackthorn, Captain McShea ordered the anchor chain cut free on the forecastle and released into the water. After waiting for the "higher" high tide, Capricorn was freed at 1920 with the assistance of tugs. She then continued under her own power to Weedon Island Station where she was unloaded.

Upon inspection Capricorn was found to have received damage in her port bow area to side shell plating, deck plating, bulwark, life rail, anchor crown & fluke plates, hawse pipe structure & bolster, and various internals in the general area. The brake for the port wildcat was burned and the port devils' claw was bent. As a result of the grounding some sections of the starboard bilge keel were torn away, and others were set up. A section of the flat plate keel was indented, and several starboard shell plates were set up with internals buckled or distorted. The tips of all propeller blades were nicked and the stern tube packing gland was leaking excessively.

LCDR George J Sepel issued a brief written statement saying "the 27 of us who survived the tragedy are praying that the rescue personnel will be able to locate other survivors. I understand my crewmen who did survive are holding up well."

At 1650 on the 30th of January the Coast Guard called off its active search for survivors, but diving operations continued.

Navy divers determined that Blackthorn was indeed resting on her port side 100° from vertical on a heading of 287°T. The divers buoyed the bow & stern. An extensive internal survey was conducted primarily to determine if any of the crew members might be trapped and alive within the confines of Blackthorn. The results were negative. However, debris inside Blackthorn restricted access to certain spaces, precluding a high confidence search of all interior spaces. The 7' by 7' cargo hold hatch was found lying on the bottom about 15' away from the buoy deck, which was in a vertical orientation.

As with the previous 2 stories, a Coast Guard Marine Board of Investigation was established and commenced on the 31st of January 1980 at MacDill Air Force Base. It consisted of Rear Admiral Norman C. Venske – District Commander in the St. Louis area, Captain Benjamin Joyce – Commander of Vessel Traffic Control in New Orleans, Captain John B. Eckman – A member of the Marine Staff Office in New York City, Commander Rene N. Roussel – Member of the District Legal Staff in San Francisco, and Lieutenant Commander John Carney Jr. – Chief of Investigation Division at the Marine Safety Office in San Diego.

The National Transportation Safety Board also joined the investigation into the collision. Their panel was made up of three civilians. As previously mentioned in this book, they are also in the Department of Transportation, along with the Coast Guard. They would eventually produce their own separate report on the collision.

A third investigation was also being put together by the State Board of Pilot Commissioners.

As the Board of Investigation got underway the operators of Capricorn, Apex Marine Corp of New York & owners, the Kingston Shipping Company filed a $1.5 million dollar damage suit blaming the Blackthorn for the collision.

And once again, as the investigation got underway, lawyers for Capricorn challenged the validity of a Coast Guard investigation into an accident involving one of its own vessels. The Capricorn attorneys demanded a panel of civilian Master Mariners to perform the inquiry and the five-member Coast Guard panel should disqualify itself. Rear Admiral Norman Venske said he would take the request under advisement. The next day, he denied the request without explanation.

As eleven of the Blackthorn survivors returned to their homeport of Galveston via a C-130 to New Orleans and bus for the final leg, they were instructed by Coast Guard officials not to speak with reporters until after the investigation.

The Coast Guard also assembled a 19-man team of divers headed up by Navy Lieutenant Commander Stephen Delaplane, who raised the Coast Guard Cutter Cuyahoga. The team will search the Blackthorn for bodies and begin salvage operations.[56]

On the 3rd of February the body of ENS Frank J. Sarna III was recovered on the surface of the water 50 yards south of Blackthorn's resting place. His cause of death was listed as drowning. He was buried at Arlington National Cemetery in Fort Myer, Virginia.

On the 4th of February Lieutenant Commander Stephen Delaplane commenced efforts to rig the Blackthorn to be raised utilizing 3 commercial heavy lift derrick barges, the Cappy Bisso (650 tons lift), Little David (100 tons lift) and Kenyon (100 tons lift).

On the 5th of February SA William Flores' body was found on the surface of the water 1 mile southwest of the Sunshine Skyway Bridge center span. His cause of death was listed as drowning. He was buried at Benbrook Cemetery in Fort Worth, Texas.

On the 6th of February QM3 Richard W. Gauld's body was recovered on the surface of the water about 1 mile south of Egmont Key, roughly 5-1/2 miles from the collision site. His cause of death was listed as drowning. He was buried in Birmingham, Alabama.

On the 18th of February LCDR Sepel testified at the Board of Investigation. He was first called on the 10th but declined to testify under Article 31. He changed his mind after hearing testimony from Capricorn. His testimony placed the blame of the collision on the Capricorn. He said he clearly saw the tanker shortly after the Blackthorn passed under the Sunshine Skyway and Blackthorn was in the center of the right-hand channel. SA Gatz had testified earlier in the investigation that he had looked up after passing under the bridge and saw the channel lights placing Blackthorn in the same location.

He went on to say that he thought his evasive action had avoided a collision, but the Capricorn kept turning to port and into Blackthorn.

On the 19th of February the Blackthorn was refloated, dewatered, and 14 missing crewmembers were discovered in the cutter. Ten were found in the engine room, five of those near the entry way, three were in a damage control party area behind the messdeck, and the 14th person was found in a chartroom behind the bridge. The Board of Investigation report unfortunately does not state exactly where all 14 individual bodies were found.

MKC Luther D. Stidhem, cause of death was listed as drowning. He was buried at Galveston Memorial Park in Galveston, Texas.

QM2 Gary Wayne Crumly was found in the chartroom. His cause of death was drowning. An autopsy report conjectured that there was evidence that he may have suffocated rather than drowned, but it was not conclusive. He was buried at Jefferson Memorial in Birmingham, Alabama.

MK2 Richard D. Boone, cause of death was listed as drowning. But his autopsy report raised the possibility of suffocation. He was buried at Lakewood Memorial Park in Hughson, California

DC2 Daniel M. Estrada, cause of death was listed as drowning. He was buried at Southlawn Cemetery in Tucson, Arizona.

GM2 Thomas R. Faulkner, cause of death was listed as drowning. He was buried at Forest Park East Cemetery in League City, Texas.

DC3 Lawrence D. Frye, cause of death was listed as drowning. He was buried at Houston National Cemetery in Houston, Texas.

SS3 Donald R. Frank, cause of death was listed as drowning. He was buried at Live Oak Memorial Park in Jefferson County, Texas.

EM3 Edward F. Sindelar III, cause of death was listed as drowning. He was buried at Arlington National Cemetery in Fort Myer, Virginia.

SNGM Randolph B. Barnaby, cause of death was listed as drowning. He was buried at sea in the Gulf of Mexico off of Galveston, Texas.

FA Michael K. Luke, cause of death was listed as drowning. But his autopsy report raised the possibility of suffocation. He was buried at Restvale Cemetery in Worth, Illinois.

SA Warren R. Brewer, cause of death was listed as drowning. He was buried at the National Cemetery in Memphis, Tennessee.

SA Glen E. Harrison, cause of death was listed as drowning. He was buried at Wilhelm-Thurston Crematory St. Petersburg, Florida.

SA John E. Prosko, cause of death was listed as drowning. He was buried at Sugar Grove Cemetery in Toronto, Ohio.

SA George R. Rovolis Jr., cause of death was listed as drowning. He was buried at Greenwich Cemetery in Savannah, Georgia.

Blackthorn being raised

On the 20th of February Blackthorn was towed back to Gulf Tampa Drydock in Tampa where she was drydocked.

After being drydocked, Blackthorn was examined to determine the extent of damage resulting from the collision, subsequent capsizing and sinking. As a result of the collision Blackthorn received extensive damage on the port side forecastle deck, bulwark, fashion plate and the sheer strake between frames 57 and 64. The superstructure between the main and 01 decks from frame 70 to 98 was torn and holed with all internals in the area extensively damaged and distorted. The port bridge wing was bent and torn adrift at the base. The port boat and davits were extensively damaged.

As a result of the sinking there was extensive damage to the machinery, equipment and entire vessel due to exposure to sea water, salt, oil, debris and marine growth. There was damage also to deck plating, shell plating, and internals caused by bottom contact and the salvage efforts to right and dewater the vessel. Deck fittings including rails, hawser reels, jackstaff mast, antennas, awnings, lights and other fittings were damaged or missing. Due to the extensive damage and the cost to make needed repairs, the Coast Guard decided to decommission and place the Blackthorn in an "inactive, out of commission in reserve" status.

Author's note - Due to the conflicting accounts of the collision, a debris field search was conducted. During Board of Investigation testimony, the Blackthorn and Capricorn crews disagreed as to the location of the collision. Capricorn was adamant that Blackthorn was on the wrong side of the channel and Blackthorn was adamant that Capricorn was on the wrong side of the channel.

There were no eyewitnesses that testified as to seeing the collision nor the exact locations of Capricorn & Blackthorn. The closest vessel was the Bayou. Her crew testified that they could see the silhouettes of Capricorn & Blackthorn and could make out sparks when they collided but could not testify as to their exact location within the channel.

DEBRIS FIELD SEARCH:

Three searches of the channel using scanning sonar were made in order to position Blackthorn and locate other debris which might be on the bottom. These scans revealed Capricorns port anchor and chain and possibly some other major pieces of debris. From there, divers were utilized to execute a detailed bottom search.

Apex Marine Corporation was contracted to provide hard hat divers and was under the supervision of 4 Coast Guard divers from the Coast Guard Atlantic Strike Team.

The Board of Investigation also utilized divers from the U.S. Navy's Explosive Ordinance Group who were under the control of LCDR Stephen Gilchrist.

The search commenced in the general vicinity of Cut "A" Channel Lighted Bell Buoy "1A" in an area based upon testimony. As soon as some debris of potential evidentiary value was detected on the bottom, a plan was developed for making a comprehensive bottom search from the line connecting Buoy "1A" to Cut "A" Channel Lighted Buoy "2A", extending eastward and westward across the width of the channel. The divers conducted this cross-channel search using two jackstays anchored on the bottom. A search was made along the jackstay, after which each end of the jackstay was moved 6 feet to the next adjacent area to be searched.

The search area extended from approximately 190' to the east and 390' to the west of Buoy "1A" and approximately 415' to the east and 210' to the west of Buoy "2A", across the channel. The probability of detection was 86%. An area of 911,000 square feet was searched. The search period extended from the 27th of February to the 25th of March.

As each item was located it was buoyed by the diver. Either the U.S. Army Corps of Engineers vessel Canaveral or Florida would pick up the buoy and plumb the buoy line to position the marked object. The position was then established by use of a Motorola Multi-Range Navigation Survey System which provided an accuracy of plus or minus 10' or the Cubic Model

40A System with an accuracy of plus or minus 2'. The degree of accuracy experienced by equipment operator Carl Nigh throughout the search was 5' or less. These positions were plotted on a U.S. Army Corps of Engineers survey chart of the area.

During the initial stages of the bottom search, the U.S. Navy divers reported that they could not relocate several pieces of debris which had been precisely positioned. The Board of Investigation evaluated the situation and requested that the Federal Bureau of Investigations consider the possibility that there might have been tampering with evidence. The FBI conducted an investigation, the results of which were inconclusive according to an informal report by the agency.

Author's note – After Navy divers surveyed the collision site, divers hired by Capricorn surveyed the site. The Navy divers later returned to the site and could not find items they had previously found.

On the 26th of March Dr. Craig Jerner of EMTEC Corp. in Norman, Oklahoma, who was a qualified metallurgical expert, visited Blackthorn while she was in drydock and compared items from the debris field. Debris from the collision site was positively matched to Blackthorn.

LOCATING THE POSITION OF COLLISION:

The Board of Investigation considered the following in evaluating the debris field:

Effect of current. This was concluded to be relatively small. For example, with a sinking rate of 5 feet per second and with a current set and drift of 060°T at 1 knot, a piece of debris could come to rest on the bottom at a position approximately 14 feet up channel or 060°T from its point of entry into the water.

Sliding effect. This was concluded to be insignificant. Pieces of debris with varying geometric shapes, which originated from adjacent areas of both vessels were found on the bottom in close proximity to each other.

Trajectory through the air. The lateral distance that debris may have traveled through the air after exiting the vessel was not taken into consideration.

Position error. The accuracy of the positioning of debris was 5 feet or less. Therefore, by combining this figure with the effects of current, the most likely position where each piece entered the water lies within an area approximately 10 by 19 feet. The channel axis 060°- 240°T, and the cross-channel axis, 150°- 330°T, intersect at the east 14 foot/5 foot point. The Board of Investigation assumed this accuracy for the critical debris, such as Capricorn's hawse pipe and Blackthorn's bulwark, when positioning the point where the pieces entered the water.

The board assumed that buoy deck pieces at frame 67 fell off Blackthorn immediately because of their proximity of other pieces on the channel bottom and the travel of Capricorn's anchor along Blackthorn's side. The nature of these pieces and their common location supports the assumption that they fell to the bottom as the collision occurred. An assumption that any of these pieces remained on the deck of Blackthorn until some later time is non-supportive. The only heeling action of Blackthorn of sufficient magnitude to drop pieces into the water was at capsizing. This location would have been in close proximity to Capricorn's anchor marks further up channel. The significant distance between Capricorn's large bolster piece and the smaller matching bolster piece indicates that they fell to the bottom close to the location of collision. The next Blackthorn debris to fall to the bottom came out as Capricorn's anchor moved aft through frames 81 and 90. It is logical to assume that the force of the anchor or chain tore out a piece of Capricorn's hawse pipe at the same time. The proximity of all these pieces on the channel bottom lends credence to the fact that they were deposited at nearly the same time.

It is important to note that pieces of Capricorn's bolster & hawse pipe form a course line that is within several degrees of Capricorn's known course prior to collision. It is believed that these pieces would not have been deflected to starboard as a result of the collision. Therefore, the center of Capricorn could not have been further north than a course line generated by these pieces. The same rationale applies to pieces of Blackthorn from frame 67 to 90. It is believed Blackthorn could not have been to the south of these pieces at the time of impact.

The Board of Investigation determined that Capricorn's bolster piece fell into Blackthorn's side during impact while both vessels were alongside each other. The rationale for this decision was that there is no other reasonable explanation for how the bolster piece could have been found on the channel bottom next to Capricorn's anchor marks. The bolster piece must have fallen out with the anchor when Blackthorn capsized.

In summary, the above discussion indicates that when Capricorn's port anchor impacted with Blackthorn, the hawse pipe was shattered, and the port side of Blackthorn was ripped so that debris from both vessels fell to the channel bottom. This sequence has been reconstructed and, when taken with the bottom positioning of the debris, accurately reflects the position of collision.

STABILITY ANALYSIS:

a. Mr. Matt Kawasaki, a stability expert from Design Associates Inc. in New Orleans, Louisiana visited Blackthorn at Luckenbach Terminal, Tampa on the 27th of March. He indicated that the two sets of marks radiating aft and down the port side of Blackthorn and in line with the aft portion of the holed side shell were made by Capricorn's anchor chain. The damaged aft portion of the port bilge keel was in way of one set of marks. Mr. Kawasaki concluded that the anchor chain exerted a force of at least 125 long tons as it damaged the bilge keel. This force was sufficient to roll Blackthorn beyond 45°, which was the point of no return for the after-damage condition.

b. Based on curves which he constructed reflecting the after-damage stability of Blackthorn, Mr. Kawasaki concluded that the maximum righting arm occurred at slightly over 20 degrees. Further, he concluded that there was probably no "down flooding" below the main deck through the damaged side shell or any other hull openings during initial capsizing because of the very short time involved. However, flooding was inevitable.

c. Mr. Christopher Loeser, a civilian naval architect from the Naval Engineering Division at Coast Guard Headquarters in Washington, D.C. evaluated the pre-damaged stability of Blackthorn. Mr. Loeser stated that the Blackthorn habitability modification and addition of a marine sanitation device were not sufficient impact to require conducting a new inclining experiment.

Author's note – Although not specified, I assume these modifications were conducted during Blackthorn's time at Gulf Tampa Shipyard.

d. Based on an estimate of Blackthorn's loaded condition prior to the collision, Mr. Loeser concluded that she met the weather criteria for small commercial cargo vessels. Also, Blackthorn met the present Coast Guard intact stability criteria. On the 28th of January the intact curves of stability for full load closely reflected Blackthorn's actual condition prior to collision. The point of no return was about 80°.

e. It was Mr. Loeser's opinion that Blackthorn could not meet a two compartment floodable length criteria.

Author's note – A stability report is provided to the Commanding Officer every evening. As the Damage Control Officer on a 210' Coast Guard cutter I was responsible for providing this liquid load & stability report to the Commanding Officer every evening at Evening Reports. This notified the CO, among other things, the maximum roll we could sustain without capsizing. On a vessel without a Damage Control Officer, this responsibility would fall to the Engineering Officer. In this case, CWO Miller.

DOCUMENTARY EVIDENCE PRESERVATION:

The FBI was also asked to see if it could restore documents, logs, and records that had been water soaked on board Blackthorn during her 22 days on the channel bottom. They were able to restore the documents and they were fully usable as evidence. Most notable was the navigation chart, which was found on the bridge crumpled up in mud & debris.

Author's note – This chart was a significant piece of evidence, and the FBI presented it in May. The Blackthorn crew hoped it would show that they were in their stated location. The fixes on the map were made by ENS Frank Sarna, who died in the collision. It showed that ENS Sarna was taking fixes every three minutes but did not mark his last fix, which would have been 2 minutes before the collision. And the markings on the chart were unclear as to Blackthorns location at the time of the collision. His previous fixes did show a steady course on the right side of the channel. But they also indicated that following that course would lead to the center off the channel.

LIFE RAFT ANALYSIS:

a. The inflatable life raft allowance for Blackthorn is four 15-man life rafts. She had her full allowance of four 15-man Navy life rafts aboard.

b. During the shipyard availability in Tampa, Blackthorn arranged through Commander, 8TH Coast Guard District for the annual inspection of her four 15-man life rafts by Atlantis Marine and Industrial Supply Co. Inc. in New Orleans, Louisiana. Both the contract and the inspection report stated that all four life rafts were Navy Mark V, whereas two were, in fact, Navy Mark III. The inspection was completed on the 5th of December 1979 and the rafts were shipped to Blackthorn at Gulf Tampa Drydock on the 6th of December, where they were stored ashore until the 27th of January.

c. Prior to stowing the four life rafts on board Blackthorn, BMC Richard Robinson removed each life raft from its soft case and unrolled it to check the equipment. The latter procedure was in accordance with instructions contained in the Coast Guard Naval Engineering Manual, CG-413. In one raft he found a 4-inch tear in an inflation chamber and two empty CO_2 bottles. He repacked the life raft without the CO_2 bottles, wrote "bad" on the flexible carrying case and stowed it in the port life raft rack on the gun tub – 02 deck. This life raft, a Navy Mark V, serial No. C6642, was on Blackthorn when she was raised. BMC Robinson also found empty CO_2 bottles in a second life raft. This life raft, a Navy Mark III, serial No. 25, was found out of its carrying case, deflated and floating in the water after Blackthorn sank.

d. BMC Robinson ascertained that the other two life rafts were in serviceable condition with full CO_2 bottles. In the process of being loaded on board Blackthorn, one of these life rafts was accidentally inflated. It was tied to the port rail of the fantail. After Blackthorn capsized and sank, this life raft, a Navy Mark III, serial No. 127, was found partially deflated and floating on the surface in a severely damaged condition.

e. The remaining good life raft, a Navy Mark V Mod I, serial No. HD1531, was stowed in the starboard life raft rack on the gun tub with the sea painter allegedly secured to a nearby rail. One of the unusable rafts was stowed on top of it. The good life raft was found on the bottom of Cut "A" Channel in its flexible carrying case by divers searching for debris, several weeks after the collision. Life raft expert Fred F. Patten of RFD-Patten Inc. in Lake Worth, Florida later inspected this lifer raft. He intentionally triggered the CO_2 activating lanyard and observed the life raft inflate fully.

f. Upon discovering that two of the life rafts were unserviceable, Blackthorn arranged to borrow a 15-man inflatable life raft, Navy Mark V, serial No. 427, from U.S.C.G.C. Steadfast (WMEC-623). It was loaded on board on the 27th of January in its rigid container and secured on the fantail. After Blackthorn capsized this life raft was found in a partially inflated condition floating upside down. This raft had last been serviced in March of 1977.

g. Blackthorn's four life rafts included two which were 24 and 25-year-old Mark III rafts. Mr. Patten stated that these rafts probably should have been replaced by newer life rafts. Although Mr. Patten would not fix a service life on inflatable life rafts, he observed that one of the Mark III's had minimum tear resistance – a condition that could only be detected by destructive testing.

h. Two of the eight CO_2 bottles inspected by Atlantis Marine were last hydrostatically tested in August of 1972. Coast Guard Naval Engineering Manual, CG-413, requires that hydrostatic tests be conducted every 5 years.

i. Coast Guard Commandant Instruction M10470.10 dated the 2nd of January 1979 refers to policies concerning the maintenance, inspection and repair of life rafts on Coast Guard vessels. This instruction pertains to Navy Mark V and Coast

Guard approved life rafts only. The instruction refers to the Naval Engineering Manual, CG-413, which further references the Naval Ship's Technical Manual for instructions concerning inflatable life rafts. Chapter 583 of the Naval Ship's Technical Manual directs that all Mark III life rafts shall be surveyed without regard to apparent condition. The reason cited is that the hull tube base cotton fabric is subject to unpredictable failure due to deterioration, resulting in a lack of reliability.

J. Chapter 583 further cautions that Mark V Mod I life raft's manual release, sea painter, should not be attached to the stowage rack. Additionally, the life raft should not be tied or lashed to the ship in any manner to eliminate the possibility of it being towed under with a sinking ship. Mr. Patten questioned the ability of any life raft in flexible containers to rise to the surface after being dragged under water. This was based upon the fact that the containers would be compressed by water pressure, thus reducing buoyancy and preventing surfacing of the raft unless the CO_2 inflation cylinders were triggered.

k. Chapter 583 states that inflatable life rafts should be located to permit ready manual overboard launching into the water without hitting any obstruction. The stowage of life rafts in the racks on the port and starboard sides of the gun tub on Blackthorn requires that each 385-pound raft be carried down a ladder to the main deck and thrown overboard. If the life raft were dropped to the next deck by releasing the hinged stowage rack, it is possible that damage to the raft would result, according to Mr. Patten.

l. Very few of Blackthorn's crew members indicated during testimony that they understood how to release and operate inflatable life rafts.

BLACKTHORN DRILLS & TRAINING

During the period from the time LCDR Sepel assumed command on the 27th of July 1979 until the collision, there were 2 "All hands" drills conducted. These drills were logged as fire drills, but LCDR Sepel testified that they progressed to the stage where abandon ship was subsequently ordered. During this period 23 duty section drills were held: 22 for fire and 1 for collision.

During the Board of Investigation proceedings, LCDR Sepel, LT Crawford, and CWO2 Miller – the Engineer & Damage Control Officer, were unable to determine basic theoretical knowledge of vessel stability. They also did not understand the information contained in the vessel's stability book.

LCDR Sepel indicated that Officers of the Deck were qualified after a period of on-the-job training, and that the procedures consisted of their being verbally questioned by qualified Officers of the Deck, the Executive Officer, and the Commanding Officer. During testimony ENS Ryan, the Conning Officer just prior to the collision, demonstrated a lack of understanding of the Inland Rules of the Road & Pilot Rules relative to whistle signals and vessels meeting in a narrow channel. As a matter of further confusion, he applied International Rules of the Road criteria in regard to whistle signals & course changes.

Author's note – There was a Course Recorder Analysis conducted on Capricorn, because she had one onboard. But it was tracking 10 minutes late so was not worth mentioning.

CONCLUSIONS:

The following are the conclusions from the Board of Investigation report finalized in May and forwarded to the Coast Guard Commandant. The Commandant released it on the 29th of December 1980.

Blackthorn and Capricorn collided at a point bearing 044°T 160 feet from the junction of Mullet Key Channel and Cut "A" Channel Range lines. This conforms closely to the location of recovered debris that first fell to the bottom, and the dead reckoning position of the Blackthorn from her last plotted fix at her computed speed of 11.7 knots.

The time of collision has been established as approximately 2021 based on the fixes on Blackthorn's chart and the time elapsed to the estimated position of impact, using a peed over the ground of 11.7 knots. This time is further corroborated by testimony, notations in Capricorn's bell book and Coast Guard Group St. Petersburg's radio log.

The headings of the two vessels at the time of impact were close to the inbound and outbound courses for Mullet Key Channel. These headings are consistent with the 180° angle of impact.

The proximate cause of the collision was that both vessels failed to keep well to that side of the channel which lay on their starboard side, when it was in fact safe and practicable to do so.

Contributing to the cause of the collision was that:

a. Blackthorn after sighting Capricorn, continued on a track of 243°T in Cut "A" Channel with her port bridge wing riding on or near the leading line and failed to use Cut "A" widener in order to provide additional sea room for the inbound vessel. This served to confuse Capricorn as to Blackthorn's intentions in the absence of radio or whistle communications.

b. Capricorn, after sighting Blackthorn, continued inbound without changing course or speed. She commenced her turn from a position slightly left of Mullet Key Channel range line into Cut "A" Channel shortly before collision. This turn, absent an outbound vessel meeting in the bend, could be considered the normal turn which would be made by an inbound deep draft vessel.

c. LCDR Sepel failed to keep appraised of the situation aboard Blackthorn and failed to effectively supervise his relatively inexperienced conning officer, particularly when Blackthorn was departing an unfamiliar port at night.

d. ENS Ryan, the Conning Officer of Blackthorn failed to immediately advise LCDR Sepel, who was on the bridge, that an inbound vessel had been sighted.

e. A port to port passing agreement was not reached due to a total reliance on radio communications which, in this case, were not successful.

f. Both Capricorn & Blackthorn failed to use whistle signals to reach a port-to-port passing agreement. This was attributed to several reasons.

First, ENS Ryan not having a complete understanding of both the use of whistle signals when approaching a bend in a narrow channel and the distinction in situations and signals between the Inland Rules & International Rules of the Road.

Second, Blackthorn's policy of not initiating whistle signals as required by Pilot Rue 80.3, if a passing agreement had been previously reached by radio.

And third, Pilot Knight's deliberate delay and subsequent failure to initiate a one short blast whistle signal on Capricorn in order to "leave the options open" to Blackthorn. If, in fact, Pilot Knight suspected that Blackthorn might proceed across his bow into the Intercostal Water Way, resulting in a crossing situation as opposed to proceeding down Mullet Key Channel, he should have initiated a one blast signal as stand-on vessel. Thus, it would appear that a one blast whistle signal would have been appropriate for either a meeting or a crossing situation even though, in the case of the latter, Capricorn would have had to alter course to the left – port, to conform to the channel.

g. Pilot Knight failed to sound the danger signal and to reduce speed as soon as he became in doubt concerning the intentions of Blackthorn.

h. Captain McShea on Capricorn failed to sound the danger signal and reduce headway when he first voiced doubt as to the intentions of the oncoming vessel, and after observing that Pilot Knight had failed to take such action.

i. Both Capricorn & Blackthorn failed to make effective use of their radar for early detection and evaluation of approaching vessels.

j. Capricorn failed to post a proper lookout by instructing the bow lookout not to report well lighted vessels in the channel. Had he reported Blackthorn upon first sighting as Kazakhstan cleared his arc of vision, advanced warning might have been given to the Captain McShea and Pilot Knight.

k. Earlier visual sightings of each vessel by the other were hindered by the brightly lighted Kazakhstan which was positioned between Capricorn and Blackthorn.

The following did NOT contribute to the collision (a through d):

a. The failure of Blackthorn to post a proper lookout, in that SA Gatz was wearing headphones over both ears, was at his post for over two hours and was neither vigilant nor properly stationed, i.e. not as far forward and low down as possible. Had an alert lookout been stationed where required, however, he would not have seen Capricorn's range lights emerge from behind Kazakhstan any sooner than personnel actually sighted Capricorn from the bridge.

b. Cut "A" Channel Range Front Light was extinguished and Mullet Key Channel Buoy "14" was off station.

c. Capricorn prior to collision had her port anchor housed with the wildcat disengaged, the brake set, the devils claw loosely secured to the anchor chain and the riding pawl disengaged.

d. Blackthorn's decision not to operate outside the channel when there was sufficient depth to do so.

The reasons for failure of attempts by Capricorn and Blackthorn to contact each other on VHF-FM Channel 13 are unknown. Blackthorn's call to Capricorn may have been overridden by Ocean Star's conversation with Kazakhstan.

Pilot Maddox on board Kazakhstan advised Pilot Knight on Capricorn prior to the collision that a Coast Guard vessel was following Kazakhstan.

Both Capricorn and Blackthorn sighted each other visually approximately 2 minutes before collision.

When Blackthorn ordered right full rudder and engines back full, collision was inevitable.

When Capricorn sounded two blasts of the whistle and put her rudder over, collision was inevitable. At the time of impact, the hard left rudder had not caused any discernable change in heading.

he proximate cause of capsizing was that Capricorn's anchor chain reached its bitter end while leading under Blackhorn's hull, came up with a strain and rolled Blackthorn to port. It cannot be determined if Blackthorn had applied an ahead bell at the time of capsizing.

The stability of Blackthorn after collision was adequate to save her had the capsizing force of Capricorn's anchor and chain not been applied.

The lack of understanding of Blackthorn's stability on the part of LCDR Sepel, Lt Crawford, and CWO2 Miller did not contribute to the capsizing of Blackthorn.

The open starboard watertight door to after berthing, 1-77-1, and the open porthole in after steering did not contribute to capsizing. By the time flooding would have occurred through these openings, Blackthorn had rolled past the point of no return due to the loss of reserve buoyancy as a result of port side damage.

The open watertight doors from the port and starboard passageways to the buoy deck, 1-70-1 and 1-70-2, the damaged port watertight door to after berthing, 1-89-2, and the open hatch, 01-135-0, from officers' country to the fantail did not contribute to capsizing and contributed very little to the rate of flooding. The effect of these open fittings was completely overshadowed by the large hole in the port side created by the collision. A tremendous inrush of water occurred as soon as Blackthorn rolled sufficiently to submerge the maindeck gunwale.

After capsizing, those crewmen aboard Blackthorn who climbed up into the engine room through the escape scuttle located in the aft messdeck area may have done so because of disorientation or limited ability to swim.

Blackthorn drifted in a direction of 059°T between the time of capsizing and the time of sinking.

Approximately 1 minute after Blackthorn capsized, Capricorn's bow grounded on a heading of 023.5°T after which she continued to pivot left – to port, due to her momentum and the flood current, finally coming to rest on a heading of 281°T 9 minutes later.

The proximate cause of the grounding was Pilot Knight's decision to maintain hard left – port, rudder following the collision and deliberately ground the vessel in order to avoid any possibility of striking the Sunshine Skyway Bridge. The turning moment generated by Capricorn's port anchor towing the Blackthorn astern, and then the anchor dragging on the channel bottom, probably made a grounding inevitable in the narrow channel.

There was insufficient all hands training on board Blackthorn in underway emergency drills since July 1979.

The inboard life raft stowage arrangement on Blackthorn's 02 deck makes compliance with the launching requirements outlined in Chapter 583, Naval Ship's Technical Manual, difficult to achieve due to the weight of the rafts.

Blackthorn did not have sufficient serviceable life rafts to accommodate all personnel. This may have contributed to the loss of a maximum of four lives.

Author's note – even though the Board of Investigation previously stated in their report that there were enough life rafts – 4. There was an issue as to whether one was actually serviceable. And an expert witness testified that 2 of the rafts were old and should not have been in service.

As far as I can tell from the crewmembers' testimony, they were only in the water a very short period of time before being rescued by the Bayou and CG41452. So, I'm not sure if the survivors would have had time to even utilize a life raft.

The immediate presence of the Bayou was responsible for the saving of many Blackthorn personnel, considering their inability to properly don lifejackets or to have life rafts available.

The fact that Blackthorn did not have one of their small boats rigged out for sea did not contribute to loss of life.

More than 25% of Capricorn's required Able Seaman were holders of U.S. Merchant Mariners Documents endorsed as Able Seaman 12-months.

No fixes were plotted on Capricorn's navigation chart during her transit from Fairway Anchorage west of Egmont Key to her grounding at Cut "A" Channel.

Merchant vessels in the Tampa Bay are not complying with the requirements of Federal Regulations regarding plotting of fixes and passing this information to the Pilot.

The impact on the environment as a result of oil pollution from this collision was negligible.

The efforts of CWO2 Miller, BM3 Bartell, HM3 Chamness, EM3 Clutter, SA Gray, and SA Rhodes with regard to post collision survival actions are commendatory and worthy of recognition.

The actions of William Parker, Vince Dyer, and Charles Whitelaw of the shrimp boat Bayou, with regard to rescue of Blackthorn survivors are commendatory and worthy of recognition.

The actions of Second Mate Stephen Sadler as boat officer and volunteers David Gilmore, Peter Hulsebosch, William Thom,

Donald Barney, Robert Rentz, and Bernie Spencer as oarsmen in the no. 4 lifeboat of Capricorn, with regard to their search efforts for survivors following the collision, are commendatory and worthy of recognition.

There is evidence of the following regarding ENS John R. Ryan, which has been forwarded to the Coast Guard for further investigation under the provisions of the Uniform Code of Military Justice:

a. Failure to sound 1 short blast of the whistle for a meeting and port to port passage. Inland Rules of the Road.

b. Failure to stay to the starboard side of mid-channel. Inland Rules of the Road.

c. Failure to sound the danger signal. Inland Rules of the Road.

d. Hazarding a vessel by not using radar to detect the presence of vessels ahead to avoid the danger of collision. Article 110, Uniform Code of Military Justice.

Failure to employ such means and devices as may be available for detecting and avoiding danger from collision. Coast Guard Regulation.

Failure to require that the lookout not remain at his post in excess of 2 hours. Coast Guard Regulation.

There is evidence of the following regarding LCDR George J. Sepel, which has been forwarded to the Coast Guard for further investigation under the provisions of the Uniform Code of Military Justice:

a. Failure to sound 1 short blast of the whistle for a meeting and port to port passage. Inland Rules of the Road.

b. Failure to stay to the starboard side of mid-channel. Inland Rules of the Road.

c. Failure to sound the danger signal. Inland Rules of the Road.

d. Hazarding a vessel by not using radar to detect the presence of vessels ahead to avoid the danger of collision. Article 110, Uniform Code of Military Justice.

e. Hazarding a vessel by failing to supervise an inexperienced conning officer in unfamiliar waters at night. Article 110, Uniform Code of Military Justice.

f. Failure to use radar when necessary for safety of the vessel. Coast Guard Regulations.

g. Failure to take special care that all precautions required by the applicable laws and regulations to prevent collisions are observed. Coast Guard Regulation.

h. Failure to ensure that all meeting and passing agreements made using radio communications are followed by appropriate whistle signals. Coast Guard Regulations.

i. Failure to station at least one lookout in the bow as far forward and as near the water as feasible when traversing a congested traffic area. Coast Guard Regulations.

j. Failure to ensure that sufficient life rafts were on board for all personnel. Coast Guard Regulations.

Failure to ensure that personnel on board were proficient in emergency drills. Coast Guard Regulations.

There is evidence of the following regarding Pilot Harry E. Knight, which has been forwarded to the Coast Guard for further investigation under Government Regulations.

a. Failure to sound 1 short blast of the whistle for a meeting and port to port passage. Inland Rules of the Road.

b. Failure to stay to the starboard side of mid-channel. Inland Rules of the Road.

c. Failure to sound 1 short blast of the whistle for a meeting and port to port passage. Inland Rules of the Road.

d. Failure to sound the danger signal when in doubt as to the intentions of Blackthorn. Inland Rules of the Road.

e. Failure to reduce speed or stop the vessel. Inland Rules of the Road.

f. Failure to evaluate the danger of each closing contact. Government Regulations.

There is evidence of the following regarding Captain George P. McShea Jr., which has been forwarded to the Coast Guard for further investigation under Government Regulations.

a. Failure to sound 1 short blast of the whistle for a meeting and port to port passage. Inland Rules of the Road.

b. Failure to stay to the starboard side of mid-channel. Inland Rules of the Road.

c. Failure to sound 1 short blast of the whistle for a meeting and port to port passage. Inland Rules of the Road.

d. Failure to sound the danger signal when in doubt as to the intentions of Blackthorn. Inland Rules of the Road.

e. Failure to reduce speed or stop the vessel. Inland Rules of the Road.

f. Failure to ensure that the position of his vessel at each fix was plotted on a chart of the area. Federal Regulations.

g. Failure to evaluate the danger of each closing contact. Federal Regulations.

i. Failure to require that no more than 25% of the required Able Seaman on board be Able Seaman 12-months. Federal Regulations.

That is the end of the Board of Investigations report. In summary they said the cause of the collision was the failure of both Capricorn and Blackthorn to keep well to the side of the channel which lay on their starboard side. But as you can see from their own testimony, which made up the narrative of the detailing of events, each captain thought they were well within their side of the channel.

There was also a report put out by the National Transportation Safety Board, which I mentioned earlier was conducting an investigation in conjunction with the Coast Guard. Their findings were similar to the point of blame. They put full blame on the Blackthorn for crossing the channel centerline into Capricorn's side of the channel and being improperly navigated.

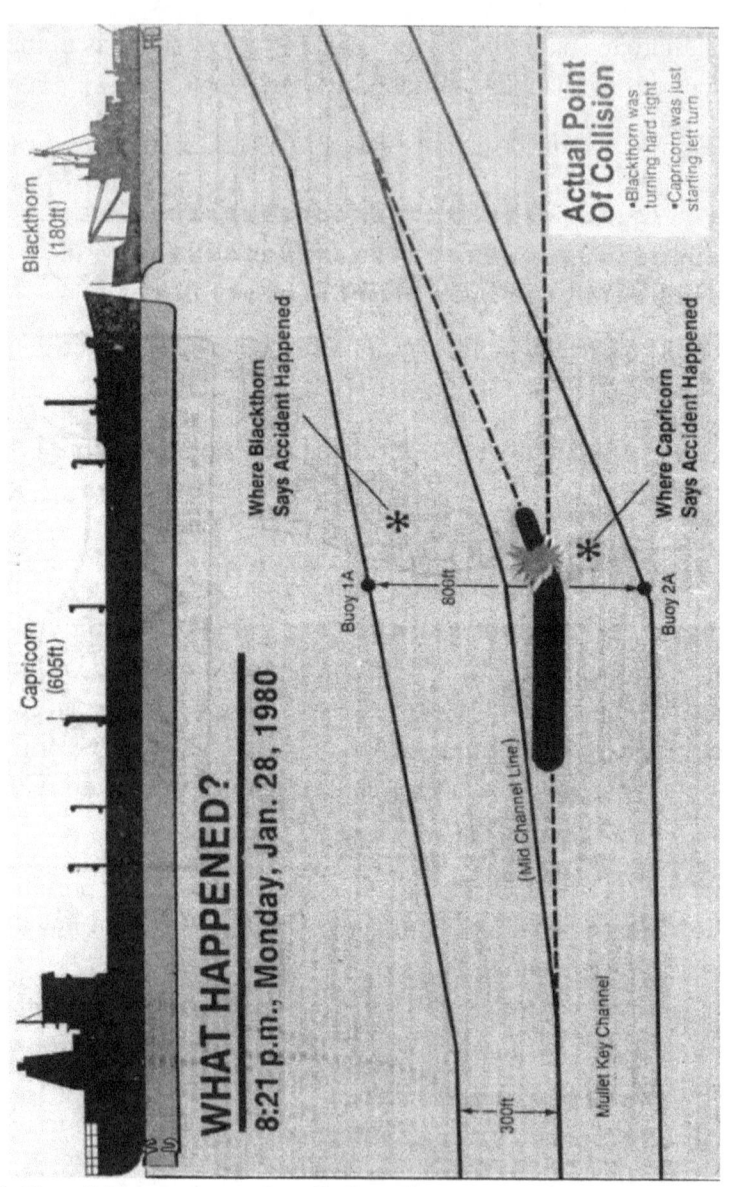

National Transportation Safety Board conclusion of where the Capricorn & Blackthorn were at point of collision[57]

In the previous picture you can see how the National Transportation Safety Board viewed the collision verse the previous drawing of how the Board of Investigation viewed the collision.

In April Blackthorn was decommissioned and after being refloated from the Gulf Tampa drydock, was moored at Luckenbach pier in Tampa. The flag aboard Blackthorn was struck by 4 survivors, LT Crawford, SN Rhodes, BM3 Bartell and FNMK Niesel.

In May 1980 CWO John S. Miller was awarded the Coast Guard Medal, the highest peacetime award the Coast Guard bestows. And BM3 Bartell and EM3 Clutter were awarded the Meritorious Service Medal. Many years later, in 2000, SA William Ray Flores was awarded the Coast Guard Medal, posthumously.

Author's note – Although absent from the Board of Investigation report, Blackthorn crewmembers lobbied to have SA Flores recognized for his heroic actions when he used his belt to hold open a lifejacket locker and remained with Blackthorn helping other shipmates. He was last seen on board Blackthorn throwing life jackets to his shipmates in the water.

In June the Coast Guard convened a board to investigate the charges recommended against LCDR Sepel and ENS Ryan. There was also a movement to investigate Captain McShea and Pilot Knight and have their licenses revoked.

No charges were ever brought against Captain McShea.

In September Pilot Knight was cleared by the State Board of Pilot Commissioners who found that no action should be taken against him regarding his state license. In April 1981 he appeared before a Coast Guard board. It wasn't until April of 1982 that a judge found him guilty of negligence.

In January of 1981, just a few weeks after the Board of Investigation Report was released, court martial charges against ENS Ryan were dropped. In April he would later receive a letter of reprimand. And in March court martial charges were dropped against LCDR Sepel and he was given a letter of admonition, which was the lightest punishment possible.

LTjg Ryan's – he was promoted since the collision, letter of reprimand was one step more severe than LCDR Sepel's letter of admonishment.

In memory of the 23 Blackthorn crewmembers who lost their lives, a memorial monument was erected on the north end of the Sunshine Skyway. The area was dedicated as Blackthorn Memorial Park on the 28th of January 1981, the one-year anniversary of the collision. The memorial was funded by a grassroots effort by the local Coast Guard Chief Petty Officers Association to raise funds and build a monument.

And on the 30th of July, 300 pounds of plastic explosives sent the decommissioned Blackthorn to the bottom once again as part of the Pinellas II Reef Site in the Gulf of Mexico.

Legal battles persisted into 1982 with finger pointing and a multimillion-dollar lawsuit as to who was at fault. Then in February a Federal Judge ruled both the Capricorn and the Blackthorn shared joint responsibility for the collision.

CLOSING

I think it is eerily coincidental how the three worst Coast Guard loss of life events since World War II could be so similar in nature. That all three events would involve Coast Guard cutters steering into much larger civilian vessels, capsizing, and sinking. And in doing so, claiming the lives of a majority of their crews.

Although I spent over 22 years in the Coast Guard as an enlisted man and officer, I had never heard of the White Alder nor the Cuyahoga tragedies. The Blackthorn I had vaguely heard of. And my knowledge of her collision was full of errors. And I'm not alone in my limited knowledge of these tragedies. While researching my book "Coast Guard Tragedies," in which I loosely touched on these three collisions, I had brought them up at Coast Guard reunions I had attended. Most of my shipmates were unaware, or had limited knowledge, of these events. So, I hope this book helps educate my fellow retired shipmates and those still on active duty.

I would have to say that with over 11 years of time at sea in the Coast Guard I never feared sinking or drowning. As an engineer I never spent much time on the bridge, but always had total confidence with those who performed their duties there. The 6 cutters I served on all had very rigorous qualifications in watch standing and damage control. Drills were usually performed daily by the duty section and at least weekly by the entire cutter underway in an all-hands scenario. Since I started my service just after the Blackthorn tragedy, I would like to think the Coast Guard had modified their qualifications and training from lessons learned from these three events. And since there has not been a loss of life incident on the scale of the White Alder, Cuyahoga, and Blackthorn since 1980, maybe the lessons learned in those tragedies have saved the lives of many a Coast Guardsmen since then.

Fair winds and following seas

REFERENCES

1. Marine Casualty Report between the U.S.C.G.C. White Alder & S.S. Helena on the 7th of December 1968.

2. The Pantograph – 9 Dec 1968, page 3

3. Tampa Bay Times 09 Dec 1968, page 7

4. The Bangor Daily News 09 Dec 1968, page 1&2

5. Stockton Evening and Sunday Record 15 Jan 1969, page 26

6. Sun Journal 04 Feb 1969, page 16

7. Cameron Pilot 13 March 1969, page 1

8. Waterwaysjournal.net

9. Dayton Daily News 19 Dec 1968, page 6

10. The White Castel Times 13 Dec 1968, page 1

11. The Shreveport Journal 13 Dec 1968, page 6

12. Vicksburg Evening Post 13 Dec 1968, page 9

13. The Town Talk 13 Dec 1968, page 16

14. Winston-Salem Journal 09 Dec 1968, page 1

15. Winston-Salem Journal 09 Dec 1968, page 16

16. Daily World 11 Dec 1969, page 1

17. The Tampa Times 14 Dec 1968, page 2

18. The Daily Adviser 07 Dec 1969, page 26

19. The Times 08 Jan 1969, page 4

20. The Tampa Bay Times 07 Mar 1969, page 24

21. The State 30 July 1969, page 15

22. The White Castle Times 05 Sept 1969, page 1

23. Themodelshipwright.com

24. Marine Casualty Report between the U.S.C.G.C. Cuyahoga & M/V Santa Cruz II on the 20th of October 1978.

25. Asbury Park Press 24 Oct 1978, page 1

26. Histamar.com.ar

27. Daily Press 06 Dec 1978, page 7

28. Daily Press 30 Oct 1978, page 8

29. Hawaii Tribune-Herald 05 Nov 1978, page 9

30. Democrat and Chronicle 22 Oct 1978, page 1

31. Richmond Times-Dispatch 09 Nov 1978, page 21

32. The Morning Call 24 October 1978, page 4

33. Fort Worth Star- Telegram 24 Oct 1978, page 6

34. The Evening Sun 16 Nov 1978, page 27

35 Richmond Times -Dispatch 09 Nov 1978, page 21

36. Press of Atlantic City 22 Oct 1978, page 2

37. The Daily Progress 22 Oct 1978, page 13

38. Richmond Times-Dispatch 27 Oct 1978, page 15

39. The Evening Sun 24 Oct 1978, page 22

40. Danville Register and Bee 26 Oct 1978, page 18

41. The Evening Sun 27 Oct 1978, page 28

42. Hawaii Tribune-Herald 05 Nov 1978, page 9

43. The Evening Sun 22 Dec 1979, page 18

44. The Evening Sun 15 Feb 1980, page 30

45. Marine Casualty Report between the U.S.C.G.C. Blackthorn & S.S. Capricorn on the 28t of January 1980.

46. aukevisser.ni

47. The Commercial Appeal 01 Feb 1980, page 4

48. The Daily Advocate 04 Feb 1980, page 1

49. The Modesto Bee 01 Feb 1980, page 1

50. The Tampa Tribune 16 Feb 1980, page 15

51. The Tampa Tribune 14 Feb 1980, page 33

52. The Maimi Herald 30 Jan 1980, page 8

53. The Tampa Tribune 27 Feb 1980, page 43

54. The Tampa Tribune 31 Jan 1980, page 2

55. The Miami Herald 30 Jan 1980, page 70

56. The Tampa Times 31 Jan 1980, page 1

57. The Tampa Tribune 01 Sept 1980, page 2

58. The Tampa Bay Times 15 Feb 1980, page 13

59. The Tampa Times 29 Jan 1980, page 1

60. Fort Worth Star-Telegram 02 Feb 1980, page 68

61. The Tampa Tribune 14 Jan 1981, page 32

ABOUT THE AUTHOR

Ed Semler retired from the United States Coast Guard in December of 2007 with over 25 years of military service in both the United States Army and United States Coast Guard. In the United States Army he was an enlisted man and was honorably discharged as a Specialist Four (E-4). While in the United States Coast Guard he was enlisted, obtaining the rank of Master Chief Petty Officer (E-9), was commissioned as an officer, and retired as a Lieutenant (O-3E).

Fully retired, he resides in Schulenburg, Texas with his wife Jana, a retired Air Force senior master sergeant. Please feel free to check out Ed's other books at www.edsemler.com, email him at mkcm378@gmail.com and check out his YouTube channel www.youtube.com/@MKCMLT

His other publications are:

"Around The World," a memoir of his 25 years of service as an officer and enlisted man in the U.S. Army and U.S. Coast Guard

"U.S. Coast Guard Cutter Sherman (WHEC-720) Circumnavigation Deployment 2001" which details the *Sherman's* historic circumnavigation of the globe and deployment to the Persian Gulf in 2001

"The Three Gunsallus Brothers" a story about fighting for Pennsylvania during the Civil War

"Sam Houston & Napoleon Bonaparte Meet On The Civil War Battlefield" a true story of the Walker brothers

"Thoughts On Being A Chief Petty Officer" a take on military leadership

"Fighting For Pennsylvania In The Early Years 1763 to 1783 – The Story of Captain Thomas Askey And Lieutenant Richard Gunsalus Of Cumberland County"

"Joe Semler Playing Baseball in the 1920's &30's"

"Alice Springs Australia Adventures In The 80's"

"Count On Us Coast Guard Cutter Dependable – Law Enforcement And Search & Rescue"

"United States Coast Guard Tragedies"

"In Their Own Words – Short Stories of Pennsylvanians in the Revolutionary War"

"American Sailors & Marines During The Revolutionary War – In Their Own Words"

"Combat Engineer Bridge Specialist"

www.ingramcontent.com/pod-product-compliance
Lightning Source LLC
LaVergne TN
LVHW051728080426
835511LV00018B/2946